Portuguese Intervention in the Manila Galleon Trade

The structure and networks of trade between Asia and America in the 16th and 17th centuries as revealed by Chinese ceramics and Spanish archives

Etsuko Miyata

Archaeopress Archaeology

Archaeopress Publishing Ltd
Gordon House
276 Banbury Road
Oxford OX2 7ED

www.archaeopress.com

ISBN 978 1 78491 532 2
ISBN 978 1 78491 533 9 (e-Pdf)

© Archaeopress and E Miyata 2016

Cover: Jingdezhen blue and white plate Fundação Anastásio Gonçalves

All rights reserved. No part of this book may be reproduced, in any form or
by any means, electronic, mechanical, photocopying or otherwise,
without the prior written permission of the copyright owners.

This book is available direct from Archaeopress or from our website www.archaeopress.com

Contents

Introduction 1
   The objectives of this study 1
   European power and Asian entrepôts in the 16th century 3
   Rivalries and dependence between Spain and Portugal in Asia 4
   The influence of Asian luxury goods on Mexican culture 4
   The trade and transport system in New Spain 5
   Veracruz to Seville 6
   Spanish markets for Asian products 6

Chapter I
The Arrival of the Portuguese and Spanish in Asian Waters 8
   1. The expansion of Iberian power towards Asia 8
   2. Trade in the Asian region before the arrival of the Portuguese and Spanish and their first contacts 9
   3. The establishment of Manila 14
   4. Trade in Manila, Macao and other port-cities 17
   5. The Chinese in Manila 19

Chapter II
Commerce and Merchants in the Manila Galleon Trade 22
   1. Flows of goods from Manila to New Spain 22
   2. Market structure and Mexican merchants 26
   3. The participation of religious orders and the issue of contraband 28
   5. The merchant diaspora and networks 30

Chapter III
Exported Chinese Porcelain in New Spain 32
   1. The export route from southern China to New Spain 32
   2. Chronology and typology of exported Chinese porcelain excavated in the city of Mexico and its change during the 16th and 17th centuries 33
      Phase 1: Mid 16th to 1575 36
      Phase 2: 1575 to the early 17th century 36
      Phase 3. First half of the 17th century 39
      Phase 4. Mid 17th century 40
      Phase 5. 1690s onwards 42
   3. Material culture and porcelain in the society of New Spain 43

Chapter IV
Distribution of Chinese Ceramics and Asian Products in Spanish Society 45
   1. Archival study of Asian products exported to Seville from Veracruz 46
   2. Chinese porcelain excavated from Seville 47
      2-1. San Juan de Acre 47
      2-2. Calle San Fernando 47
      2-3. Real Monasterio de Santa Clara 47
      2-4. Altamira 48
      2-5. La Florida 48
      2-6. Cuartel del Carmen 48
      2-7. Date phases 48
         The mid 16th century to 1575 48
         1575 to the late 16th century 49
         The early 17th century 50
         The late 17th to the early 18th century 50
         18th-century pieces 50
      2-8. Some short remarks 53
   3. Chinese Porcelain in Lisbon and the Galician Coast 54
      3-1 Excavated ceramics from Bayona 54
      3-2 Excavated ceramics from Vigo 57

i

    3-3. Excavated ceramics from Santiago de Compostela 58
    3-4. Excavated ceramics from Orense 60
    3-5. Significance of Chinese ceramics in the Atlantic coastal trade 61
  4. Classification of ceramics from Lisbon 61
    4-1. The earliest type of ceramics 61
    4-2. The first half of the 16th century 62
    4-3. The mid 16th century to early Wanli period 63
    4-4. The third quarter of the 16th century 64
  5. The porcelain trade from Lisbon to Amsterdam 66
  6. Conclusion 68

Glossary 70
  Chinese Dynasties and Periods 70
  Unpublished Materials 71

Bibliography 71
  Published Materials 72
  Catalogues 75

Appendix 1
AGN Contratación 1795-1802 76

List of Figures, Tables and Photographs

Figure 1: Model of trade network 2

Photograph 1: Chinese ceramic sherds excavated from Alfama, Lisbon.(©Museu da Cidade, Lisboa) 11

Photograph 2: Jingdezhen blue-and-white ewer dated to the early 16th century.(©Fundação Alameida) 12

Photograph 3: Jingdezhen large blue-and-white basin dated to the middle of the 16th century (excavated from Templo Mayor, Mexico City, © INAH). 15

Photograph 4: Jingdezhen large blue-and-white bowl dated to middle of the 16th century (excavated from Templo Mayor, Mexico City, © INAH). 15

Photograph 5: Jingdezhen large blue-and-white bowl dated to middle of the 16th century (excavated from Templo Mayor, Mexico City, © INAH). 15

Map 1: Plan of Manila in 1671 ©AGI Filipinas, MP10. 16

Numbers of *sangleyes* entering and leaving Manila in 1604. 19

Figure 2: Taxes paid by Mexican merchants. 28

Figure 3: Number of merchants participating in trans-Pacific trade. 28

Photograph 6: Jingdezhen polychrome bowl with double-headed eagle from the excavations at La Calle Licenciado de Verdad ©INAH..... 29

Figure 3: List of confiscated contraband.. 30

Map 2: Map of excavated area of Zócalo (©INAH)... 33

Figure 4: Original production sites of ceramics found in Mexico City... 34

Photograph 7: *Kraak* porcelain plate recovered from the *Witte Leeuw* © Rijksmuseum 34

Photograph 8: Jingdezhen blue-and-white sherd excavated from the Templo Mayor, Mexico City © INAH. 36

Photograph 9: Jingdezhen blue and white plate © Fundação Almeida 37

Photograph 10: Jingdezhen blue-and-white sherds with vegetal motif excavated from the Donceles Street site, Mexico City (© INAH). 37

Photograph 11: Jingdezhen blue-and-white sherds excavated from the Templo Mayor, Mexico City (© INAH). 37

Photograph 12: Jingdezhen blue-and-white sherds excavated from the Templo Mayor, Mexico City (© INAH). 37

Photograph 13: Jingdezhen blue-and-white plate with phoenix motif excavated from the Templo Mayor, Mexico City (© INAH). 38

Photograph 14: Jingdezhen blue-and-white sherds of plates from the Templo Mayor, Mexico City (© INAH). 38

Photograph 15: Jingdezhen blue and white plate Fundação Anastásio Gonçalves 38

Photograph 16: Jingdezhen blue-and-white sherds from *Nossa Senhora dos Máritires* (©CNAS-DGPC). 39

Photograph 17: Jingdezhen blue and white sherds of *kraak* bowls excavated from the Templo Mayor, Mexico City (© INAH). .. 39

Photograph 18: Jingdezhen blue and white sherd excavated from the Templo Mayor, Mexico City (© INAH). 39

Photograph 19: Jingdezhen blue-and-white sherds excavated from the Cathedral site (Zócalo, Mexico City-© INAH)................. 40

Photograph 20: Jingdezhen blue-and-white sherd excavated from Zócalo, Mexico City (© INAH)........................ 40

Photograph 21: Jingdezhen blue and white sherd excavated from the Donceles Street, Mexico City (© INAH)........................ 41

Photograph 22: Jingdezhen blue and white sherds excavated from the Donceles Street excavations, Mexico City (© INAH)...... 41

Photograph 23: Jingdezhen blue and white sherds of "crow cup" excavated from Zócalo, Mexico City (© INAH)........................ 41

Photograph 24: Jingdezhen blue-and-white sherds excavated from the Templo Mayor, Mexico City (left). Jingdezhen blue and white sherds excavated from La Calle Licenciado Verdad (right) (© INAH). 42

Figure 5: Chronological data based on the Chinese ceramics found at the Zócalo área, Mexico City.. 42

Photograph 25: Wicker basket containing Japanese ceramics (©Amsterdam City Museum)................................ 43

Photograph 26: Gilded Jingdezhen polychrome vase decorated in ex-private residence in Zócalo, currently a restaurant (©INAH) 44

Map 3 Actual map of historical part of Seville ©Junta de Andalucía. 47

Photograph 27: Jingdezhen blue-and-white bottle excavated from the Real Monasterio de Santa Clara, Seville, dated c. 1570-1575 (below left) (© Museo Arqueológico de Sevilla). 48

Photograph 28: Jingdezhen blue-and-white bowl with bird design dated to c. 1550-1575, excavated from Cuartel del Carmen, Seville (© Museo Arqueológico de Sevilla). 49

Photograph 29: Jingdezhen blue-and-white sherds of plates excavated from Cuartel del Carmen(© Museo Arqueológico de Sevilla). ... 49

Photograph 30: Jingdezhen blue-and-white plate dated from 1575 to the late 16th century excavated from San Juan (© Museo Arqueológico de Sevilla). ... 49

Photograph 31: Jingdezhen blue-and-white sherds of plate with taihu rockery design excavated from Calle San Fernando (© Museo Arqueológico de Sevilla) ... 50

Photograph 32: Jingdezhen blue-and-white sherd of bottle excavated from San Juan de Acre, Seville (© Museo Arqueológico de Sevilla) ... 50

Photograph 33: Jingdezhen blue-and-white sherd of tall cup excavated from San Juan de Acre, Seville (© Museo Arqueológico de Sevilla). ... 51

Photograph 34: Possibly Guangdong (?) blue-and-white sherd of plate with flower-basket motif excavated from San Juan de Acre, Seville (© Museo Arqueológico de Sevilla) ... 51

Photograph 35: Dehua blue-and-white cup excavated from San Juan de Acre, Seville (© Museo Arqueológico de Sevilla) ... 52

Photograph 36: Jingdezhen polychrome bowl with cover found at the La Florida site, Seville (© Museo Arqueológico de Sevilla). ... 52

Figure 6. Quantity of excavated Chinese porcelain in Seville ... 52

Photograph 37: Seville blue-and-white ware with Chinese influence (left) (© Museo Arqueológico de Sevilla). Jingdezhen blue and white plate from the *Vung Thau* (right) (© Museo Nacional del Arte Decorativo). ... 53

Map 4: Map of Spain and Portugal. ... 54

Photograph 38: Jingdezhen blue-and-white sherd of plate excavated from Monterreal.©Museo do Mar ... 55

Photograph 39: Jingdezhen blue-and-white plate with *linzhi* motif on the rim excavated from Monterreal ... 55

Photograph 40: Jingdezhen blue-and-white sherd of bowl with egret and floral design excavated from Monterreal. ... 56

Photograph 41: Jingdezhen blue-and-white sherd of plate with deer and pine motif excavated from Monterreal. ... 56

Photograph 42: Jingdezhen blue-and-white sherd of plate with egret motif excavated from Monterreal. ... 56

Photograph 43: Jingdezhen blue-and-white sherd of plate with deer design excavated from Monterreal. ... 57

Photograph 44: Jingdezhen blue-and-white sherd of plate excavated from Calle Hospital, Vigo. ... 57

Photograph 45: Jingdezhen blue-and-white sherds of bottle dated to the mid 16th century excavated from Casa del Deán. ... 58

Photograph 46: Jingdezhen blue-and-white sherd of plate excavated from Casa del Deán. ... 59

Photograph 47: Jingdezhen blue-and-white sherd of deep bowl (*klapmutsen*) excavated from Casa del Deán ... 59

Photograph 48: Large Jingdezhen blue-and-white *kraak*-type moulded plate excavated from Casa del Deán. ... 60

Photograph 49: Jingdezhen blue-and-white sherd of plate excavated from Casa del Deán. ... 60

Photograph 50: Jingdezhen blue-and-white sherds of plates excavated from the Maceda site. ... 60

Map 5: Lisbon. ... 62

Photograph 51: Chinese ceramic sherds excavated from Alfama (© Museu da Cidade). ... 62

Photograph 52: Chinese ceramic sherd with conch shell motif in the centre excavated from Alfama (below right) (© Museu da Cidade)... 63

Photograph 53: Blue-and-white bowl with floral spray design excavated from Alfama (left) (© Museu da Cidade) ... 63

Photograph 54: Chinese ceramics with cross mark on the base excavated from Alfama (© Museu da Cidade) ... 63

Photograph 55: Chinese ceramics with reign marks on the base excavated from Alfama(© Museu da Cidade). ... 64

Photograph 56: Jingdezhen blue-and-white large dish with wave design on the interior and scroll design on the exterior excavated from Alfama (© Museu da Cidade). ... 64

Photograph 57: Jingdezhen blue-and-white large plates excavated from Alfama(© Museu da Cidade). ... 65

Photograph 58: Jingdezhen blue-and-white sherd of small bowl with circular motif on the exterior excavated from Alfama(© Museu da Cidade). ... 65

Photograph 59: Jingdezhen blue-and-white sherds of bowls excavated from Alfama(© Museu da Cidade). ... 65

Photograph 60: Jingdezhen blue-and-white sherds of bowls generally known as *wantouxin* excavated from Alfama (© Museu da Cidade). ... 66

Photograph 61: Jingdezhen blue-and-white plate with phoenix motif on the central medallion excavated from Alfama (© Museu da Cidade). ... 66

Introduction

The 16th-century world economy experienced a diverse change in many regions, the key products being silver, sugar, spices, slaves and silk. With the arrival of the Portuguese in Asian waters in the early 16th century, the regions began exporting to Europe spices (cloves, peppers) and other luxury goods, such as silk, Japanese silver and Chinese ceramics.[1] The Portuguese also exported slaves from Africa to Europe and America, and sugar from Brazil. This sugar/slave trade created a new economic cycle, exploiting sugar as a 'global' commodity. Additionally, after the conquest of America, Spanish interests began exporting great quantities of American silver to Europe and Asia, and brought silks and other Asian goods to America in return. By the 16th century silver was an important component of the Asian economy and before the beginning of the Manila galleon trade (1565), Japan had major interests in the supply of silver to Asia. The shortage of silver in the Chinese economy in the 16th century was the key driver of silk exports, in order to acquire silver, through trade with the Portuguese and Spanish. Large quantities of silk were traded in south China, Macao, Formosa, Nagasaki and Manila.

The objectives of this study

This work will study closely the trade structure and function of the Manila galleon trade, especially during the 16th and 17th centuries, in Asia and New Spain. The trading of silk for silver has been so exaggerated in terms of Asian and American trade that it is often overlooked that it also had a major influence on the transportation of other goods, peoples, cultures, and the actual merchants who participated in the trade between the two different regions. One core aim of this study is to propose the importance of the participation of Portuguese merchants in the Manila galleon trade in Asia using their extended network and their connections with New Spain, a factor which seems never to have been discussed in the past. Portuguese trade between Macao, Manila and Nagasaki has been studied by Charles Boxer,[2] Pierre Chaunu[3] and Manel Ollé.[4] Yoshitomo Okamoto[5] and Koichiro Takase[6] have carried out important studies on the trading relationships of Japan and Macao and the intervention of the Jesuits in the circulation of slaves, silk and silver.

The Macao–Manila–Mexico route was important as a trunk network for Pacific trade, since Chinese and other Asian products to be shipped were first obtained in Guangdong and Fujian and then brought to Manila via Macao, or directly from southern China. Rivalry between Portuguese and Chinese merchants is often discussed, although supply from Macao, especially from early dealings until the prosperous period of the Manila galleon trade, was very important to the Spanish because of their lack of commercial and political knowledge and their connections to other Asian and Southeast Asian countries.

Many of the powerful merchants participating in this trade were Portuguese *conversos*,[7] with their networks through Asia, America and Europe. These firms were especially important for their links with Mexican merchants and the Manila galleon trade. Although not all of these traders were successful in the new territories escaping from Portuguese controls, what differentiated them from other Portuguese, Spanish or Chinese merchants was that they had an extended network, being Jews, and thus were able to work as merchants in every location, with their local slaves, the trade in which they were actively engaged. The importance of Portuguese *converso* merchants in Mexican society has been studied by Jonathan I. Israel[8] and Louisa Hoberman;[9] the trading activities of these *conversos* on the Asian side has been partly studied by Lucio de Sousa.[10] The Spanish merchant structure in Mexico (and its system in relation to trans-Pacific trade) has also been investigated by Carmen Yuste,[11] Carlos Martínez Shaw[12] and, of course, in the work of William Lytle Shurz.[13] All the relevant historical facts were partially explored by these scholars, although a dedicated study integrating and connecting all these regions, social elements, traded products, as well as the

---

[1] Jingdezhen wares are mostly of 'porcelain' quality, although the author will use the term 'ceramic' so as to avoid confusion.
[2] Charles R.Boxer, *O Grande Navio de Amacao*. Fundacao Oriente, Macau, 1989.
[3] Pierre Chaunu, *Les Philippines et le Pacifique del Ibérics(XVI, XVII, XVIII siécles)*, S.E.V.P.E.N, París, 1960, pp. 204-206.
[4] Manel Ollé Rodríguez, 'Macao-Manila Interactions in Ming Dynasty', *Macau During the Ming Dynasty*, Centro Cientifico e Cultural de Macau, I.P, 2009, pp. 152-176.
[5] Yoshitomo Okamoto, *Jurokuseiki Nichiou Koutsushi no Kenkyu*, Hara Shobo, Tokyo, 1974, pp.140-260.
[6] Koichiro Takase, Kirishitan Kyoukai no Boueki Katsudou-Tokuni Kiito Igai no Shouhin nitsuite, *The Socio-Economic History Society* 43 (1), Tokyo, 1977, pp. 54-72.
[7] *Conversos* is a term used for those Jews who escaped the Spanish Inquisition in 1391. They fled first to Portugal, and again migrated to America when the Portuguese Inquisition began in 1536, keeping their Jewish faith within their families and close friends in secrecy.
[8] Jonathan I. Israel, *Razas, clases sociales y vida política en el México colonial 1610-1670*, Fondo de Cultura económica, México, D.F. 1980, pp. 116-136.
[9] Luisa Schell Hoberman, *Mexico's Merchant Elite, 1590-1660. Silver, State, and Society*, Duke University Press, Durham and London, 1991, pp. 4-10.
[10] Lúcio de Sousa, 'Legal and Clandestine Trade in the History of Early Macao: Captain Landeiro, the Jewish King of the Portuguese from Macao', *Kanagawa Prefectural Institute of Language and Culture Studies*, 2012, pp. 49-63.
[11] Carmen Yuste, *El comercio de Nueva España con Filipinas 1590-1785*, Instituto Nacional de Antropología e Historia, México, 1984.
[12] Carlos Martínez Shaw (ed.), *El Pacífico español de Magallanes a Malaspina*, Madrid, Ministerio de Asuntos Exteriores y Lunwerg Editores, 1988.
[13] William Lytle Shurz, *The Manila Galleon*, E. P. Dutton, New York, 1939.

FIGURE 1: MODEL OF TRADE NETWORK

routes through Southeast Asia, southern China, Macao, Manila, Acapulco, Mexico, Veracruz, Seville (and indirect connections to the Atlantic coastal trade) has never been carried out.

The significance of this study crucially stands on this point: that all short-distance regional trades are connected to larger-scale trades, which in turn are again distributed in regional markets that further extend from the destinations of such large-scale trades (see Fig. 1).

The Manila galleon trade, using Pacific routes, constructed a complex trade network that extended and criss-crossed itself like a vast spider's web. This worldwide trade network was operated by the Portuguese *conversos* with their well-developed connections from Asia to Europe. No other commercial network in the past had been carried out in a similar way by one single ethnic diaspora, and, therefore, it is possible to trace their communities and influence by studying genealogies, friendships, and, occasionally, even specific products, for example Chinese ceramics in which they had specific rights.

This present study will also focus on trade relations between Manila and Macao, which had always been maintained, although any entrance from Manila to Macao by the Spanish was officially prohibited. In a way, Manila relied upon daily supplies brought by the Portuguese and Chinese merchants. In addition, after Japan closed the borders and prohibited trade with the Spanish and Portuguese, the latter were obliged to obtain American silver from Manila to purchase Chinese goods.

Chapter I will include a discussion on the Asian market structure and the participation of Iberian powers in this regional trade. The historical background, which made the Portuguese and Spanish the first among the Europeans to expand their powers into Asia, must be explained in order to understand the cause and effect of the new century of maritime trade in Asia. The rapid growth of Portuguese commercial activities was the key to the conversion of Manila into an international port polity at the beginning of Manila galleon trade. The Spanish used the commercial network established by the Portuguese, utilizing the existing Asian trade network and introducing many ships into Cavite to distribute American silver. The participation of Chinese merchants was always one of the major factors of Asian trade, but Chapter I will clarify that, in the case of Manila, the majority of Chinese immigrants earned their livelihoods as labourers and craftsmen rather than from trade, and that the true merchants supporting the Manila galleon trade were highly likely to have been the Portuguese.

Chapter II will explain what products were actually exported to New Spain. Silk and other textiles and ceramics of course, but the quantities of ceramics have always been vague in previous studies, and some of the statistics taken from Blair and Robertson, and used by many scholars, are unreliable, as the original sources have not been indicated. Other products and commodities, such as slaves, were almost never mentioned in previous studies, but it is very important to consider the early Asian migrations into America and the Caribbean areas, along with their cultural impact on local societies. The participation of religious orders in this trade is also mentioned in Chapter II, linking to the following chapter on the fact that many Chinese ceramics have been excavated from monasteries and convents. The nature of Mexican merchants and the importance of Portuguese *converso* merchants, who made their fortunes trading Asian goods and other products, is a central issue to be discussed here, along with their wide supporting network behind them. The connection of *conversos* to New Spain will explained how and why they were important to the Mexican economy and society.

Chapter III is dedicated to the ceramic trade routes from southern China to Macao and on to New Spain. This includes studies on some of the wreck sites, and shows how Chinese and Portuguese merchants segregated their trade products in an attempt to co-exist in the China–Manila trade, rather than creating conflict and rivalry. Further discussion is mainly given to the Chinese ceramics found in New Spain, analyzing how the supply and demand changed over the course of time and considering why the zenith of ceramic exports was in the late 16th and early 17th centuries, and also explaining the absence of Chinese ceramics in the mid-17th century. How these ceramics were accepted among the societies of New Spain, and were involved in its original development will also be discussed in order to understand the impact that Asian products had on the material culture.

Chapter IV will look at the excavated material from Spain, Portugal and the Netherlands. This archaeological study indicates the differences in material culture among these leading European nations in the 16th and 17th centuries. The datings, quantities and types of Chinese ceramics from each site demonstrate the mercantile structure, route and likely trading volume for each period. The acceptance of Chinese ceramics by Spanish and Portuguese societies, and their further distribution throughout Europe, can be differentiated by analyses of the Chinese ceramics excavated. Asian products, such as silk and ceramics, were highly valued in Europe from the previous period and were traded by Arab merchants. After the conquest of Malacca in the early 16th century, these products were directly traded and transported by Portuguese merchants to Lisbon. The nature of Asian goods were converted from being a desirable objects for the upper class aristocrats into consumer items of middle class merchants. This helps provide an overview of trade and material culture in Europe in the 16th and 17th centuries, just prior to the 'boom' in Oriental goods of the 18th century.

European power and Asian entrepôts in the 16th century

The beginning of trade between Manila and Acapulco, the longest maritime trading distance in history, changed Manila into one of the most important entrepôts in Asia. The island of Mindoro appears in a Chinese document already in the 13th century as '*Ma-I*' (麻逸) in '*Zhufanzhi*' (諸蕃誌). However, the nature of the Asian ports of the 16th century was very different from that of earlier periods following the arrival of European powers into Asian waters. Macao was officially permitted by the Chinese authority to undertake trade between Portuguese and Chinese merchants in 1557. Although it never functioned as a trading port in earlier periods, Macao became an important base for the Portuguese to acquire silks, damasks and other silk-derived goods from Guangdong. Winter-related goods from Guangdong provided opportunities for exports to India, Commodities such as silk, gold, musk, ceramics, rouge and rhubarb were brought down-river to Macao from Guangdong in large vessels known as *lanteas*, although considerable quantities of goods were also smuggled into Macao to be sold clandestinely.[14] Formosa was also a newly-established port city. Chinese from the mainland began to settle gradually there from the 1200s, but it was originally a small harbour where fishermen came to barter goods until the Fujianese and Portuguese began to call in to trade. After failing to capture Macao, the Fujianese pirates established themselves along the southwest coast of Formosa and fortified the port, which became a distribution destination for Chinese silks, Japanese silver and copper.[15] Nagasaki was another small fishing village (of only 100 households) until 1570, when Omura Sumitada (大村純忠) established a port open to Portuguese shipping. These were newly-established ports connected to Portuguese trade, where Chinese silk and Japanese silver were mainly bought and sold by Portuguese, Chinese and Japanese merchants. The Portuguese played an important role as middlemen in this trade, buying Japanese silver with Chinese silk. Among the circumstances enabling this was the fact that official trade between China and Japan was prohibited by the activities of the local 'warrior merchants' (*Wako*) along the coastal area. Additional factors included the large demand for Japanese silver in China and the growing appetite for Chinese silk in the Japanese market.[16] The majority of Chinese silks exported from Macao went to Nagasaki, although a significant quantity was left for supply to India, Europe and Spanish America (via Manila).

Conversely, Manila had already existed as a trading port in earlier periods. In the Ming dynasty the Sino-Filipino relationship was considerably developed, both commercially and economically, and almost all the trading ports were visited by Chinese merchants. Envoys from the Philippines travelled to China and paid tribute to the Ming court.[17] After the establishment of Spanish interests in this city in 1571, Manila became one of the most important ports in all Southeast Asia, where merchants from New Spain, south China, Portuguese from Macao as well as merchants from India came and traded. The nature of Manila as a port was mostly focused on acquiring American silver, however various products from other Asian bases entered the port of Cavite. Direct contact between Chinese merchants and the Spanish in Manila began around 1572: six Chinese ships arrived in Cavite in 1574, 12-15 in 1575, and 40-60 ships in 1580. The major increase in numbers of Chinese ships entering Manila can be confirmed over a short period of time.[18] *Las Cartas de Indias* (Letters from India), written between 1588 and 1591, mentions that merchants from many places, including Japan, Macao and Siam, came to trade silks, black and coloured damasks, embroideries and other textiles, *inter alia* large quantities of black and white cotton cloth. In return they brought beeswax, cotton, wood and shells, which in some places were used for money.[19] Another document also refers to active trade between Manila and China, with the major items brought from China to the Philippines being ceramics, jars, iron, silks, fine ceramics, mercury, gunpowder, peppers, cinnamon, cloves, sugar, copper, oranges, rice, gold powder, wax, etc.[20] When the Manila

---

[14] Charles R. Boxer, *The Great Ships from Amacon. Annales of Macao and the Old Japan Trade, 1555-1640*, Centro de Estudios Históricos Ultramarinos, Lisboa, 1959, pp. 5-10.
[15] Charles R. Boxer. *op. cit*, pp. 1-5.
[16] Charles R. Boxer. *op. cit*, p. 1.
[17] *Chinese in the Philippines*, Philippine Social Science and Humanities Review vol. XXIX Sep-Dec. 1964, no.3-4, pp.315-323.
[18] William Lytle Shurtz, *op. cit*, pp.72-73.
[19] Blair, Emma Helen, and Robertson, James Alexander, (eds.) *The Philippine Islands*; 1492-1898, Vol. VII, Cleveland, 1903-1909, p.35.
[20] AGI México 19-82, *Carta de Martín Enriquez a SM*, fol. 3.

galleon trade began, these Asian products were traded for American silver and exported to Acapulco. This trade transformed Manila into a port where silk and Asian luxury goods were concentrated and into a centre for silver distribution into Asia. The city was said to be highly cosmopolitan, with Japanese, Malays, Javanese, Italians, French, Greeks, Portuguese, and a large number of Chinese merchants residing in the city.[21]

Rivalries and dependence between Spain and Portugal in Asia

The great difference between Portuguese and Spanish trade in Asia was that the Spanish rarely intervened in Asian commerce, and never in the same way as the Portuguese. As previously mentioned, they were middlemen between southern China, Nagasaki, and Manila. Even the wealthy Cantonese merchants shipped their goods to Manila and Nagasaki via Portuguese merchants.[22] Silk and silver in Asia circulated between Guangdong, Macao, Formosa, Nagasaki, and Manila, with the merchants connecting these ports being Portuguese and Chinese. Chinese merchants provided silk and other textiles, whereas the Portuguese offered Japanese and American silver and silk. The trade link between Manila and Macao was quite open, and Portuguese merchants frequently brought Chinese goods acquired in the markets of Guangdong to Manila, to be sold to the Spanish at prices higher than those of Chinese merchants. However, the taxes paid by Portuguese merchants in Manila were far less than Chinese merchants had to pay, and this was complained of by the Spanish on many occasions. The relationship between the Portuguese and the Spanish in Asia was barely affected by the politics of the Iberian Peninsula, and, although the two kingdoms were united until 1640, the Portuguese had to pay taxes when trading in Manila, while the Spanish were officially prohibited from travelling to Macao from Manila. Such rivalry between two countries was prominent in Asia. On the other hand, the Portuguese authorities often requested military support from the Spanish government against Ternate, and also economic support for Formosa. On the other side, the existence of Portuguese merchants in Asia offered fundamental support for the Spanish in the Philippines. Spain needed Portuguese merchants, the latter being much more integrated and better informed in terms of Asian trade, and thus used them as intermediaries to acquire goods and participate generally in trade matters. On the other hand the Portuguese government, which did not have a stable base in Southeast Asia, and which continuously faced armed conflicts, especially from the Dutch, depended on Spain for soldiers and money. Portuguese merchants needed silver to continue developing trade with China and other countries in Southeast Asia (especially after Japan gradually closed the country off in the first decades of the 17th century), and they endeavoured to reinforce the link with Manila to acquire American silver, even sending ships directly to New Spain from Macao.

The intervention of the European powers in the Asian region beginning from the 16th century resulted effectively in redrawing the map of Asia and establishing new port-cities such as Macao, Nagasaki and Formosa, trading silver and large quantities of silk and other textile exports from China to other Asian ports, some destined for the New World and European markets. Conversely, this intervention meant a direct connection in terms of the world economy between Asia, Europe and the New World, transporting goods from Asia to New Spain, crossing the Pacific, or to Lisbon via The Cape of Good Hope. People, various products and cultures were more closely connected between Europe and Asia than before.

The influence of Asian luxury goods on Mexican culture

Asian products (silks, ceramics, lacquer wares, cotton and other textiles, furniture, wax, etc.) were well accepted all over the New World, from Mexico to Peru and other Spanish colonies. Silk was exported in large quantities from Manila to Acapulco, and from there distributed to all the other countries and cities of the continent. (In many documents silks appear as *ropas chinas* (clothes from China).[23] Silk clothes from China were quite common and were acquired, probably at reasonable prices, by middle-class Mexican society. In his book, Thomas Gage refers many times to women wearing Chinese silk and shops selling the fabric.[24] The Parián, located in the Plaza Mayor, was a market where all Asian goods were sold. Chinese porcelain, in particular, was highly prized by the wealthier classes of American society. Great quantities of ceramics, dating from the 16th to the 18th centuries, have been excavated from the wider vicinity of Zócalo. These ceramics were bought and used in monasteries, convents, and residences of rich families, where wealth was naturally concentrated.[25] Plates and bowls are the shapes most commonly found from excavation sites, although some forms, such as jars or tall cups for coffee or chocolate, are also found, helping to characterize the material culture of Mexican society at the time.

---

[21] Manel Ollé Rodríguez, *Macao-Manila Interactions in the Ming Dynasty*, Macau During the Ming Dynasty, Centro Cientifico e Cultural de Macau, I. P, 2009, pp.152-176.
[22] Charles R. Boxer. *op. cit.* p. 12.
[23] It is unclear from the original Spanish documents if silk was actually woven in China and exported as items of clothing.
[24] Thomas Gage, *Nuevo Reconocimiento de las Indias Occidentales*, 1ª edición, 1648, Fondo Cultura Económica, México, D.F., 1982, pp. 178-180.
[25] Some Chinese ceramics are noted as heirlooms in wealthy Mexican homes. Antonio Rubial García (coordinador), *Historia de la Vida Cotidiana en México, Tomo II: La Ciudad Barroca*, El Colegio de México y Fondo de Cultura Económica, México, 2005, p. 94.

Chinese ceramics also had a great influence over Talavera wares, with many of the designs and motifs deriving from the Jingdezhen repertoire. The *biombo*, a Japanese folding screen (*biobu* in Japanese), was also an item very much favoured among rich Mexicans, and the technique was copied to produce a local version of the product, featuring the scenery of Mexico city.[26] Lacquer wares also became very popular, and the inlaid shell technique of *raden* was later copied by the Mexicans and reproduced as *enconchado*, a wooden art form featuring shell inlay.[27]

The trade and transport system in New Spain

Asian products were acquired in Manila and transported to Acapulco; from there they went overland to Lima or Mexico. When the galleon reached the coastal area of Baja California, ships known as *barcos de aviso* (ships sent to Acapulco in order to notify arrivals) were dispatched to inform the port of Acapulco.[28] When the galleon docked the market (*feria*) opened and many items were sold. This market continued for a month and a half, starting from January until February. Before opening this market, the government inspected the cargo in order to confirm if the cargo register in Acapulco matched that from Cavite. This inspection was to put a limit on the cargo tonnage and prohibit contraband, although in many cases the tonnage was falsified and more goods than initially registered were normally loaded. Moreover, in order to register goods at lower prices, silk and other products of less quality were placed above the higher quality silk. On arrival at Acapulco the price of each product was reviewed and fixed among the merchants of New Spain and the Philippines.[29] Once the price was agreed upon the market in Acapulco was opened to the public, although the major part of the cargo was transported directly to Mexico. Trade between Acapulco and Peru was prohibited in 1582 so as to stop Peruvian silver being excessively transported towards Asia via Acapulco, nevertheless smuggling was always active and the flow of goods between the two regions never ceased. Peruvian merchants travelled to Acapulco when the galleon entered the port twice a year. The Peruvian economy heavily depended on silver exports and brought great quantities of silver to Mexico. Basic commodities were traded and Asian products and Peru–Mexico trade continued to represent commercial connections in America through the colonial period. What is interesting is that these Asian products were not only distributed in determined places, but can be found in almost all locations along the Caribbean coastal areas, e.g. Florida, Panama, Guatemala, Cuba and Puerto Rico.[30]

What this indicates is that the material culture of the New World was affected by the Manila galleon trade, and Asian luxury goods were in demand in almost every American settlement of any size among the wealthy classes. Those Asian products appreciated in important colonial cities, such as Mexico, Lima and Havana, were also in demand by almost all societies.

Cargoes destined for Mexico City were reloaded on mule caravans in Acapulco and headed towards the capital. The caravans normally had contracts with those merchants who were the cargo owners and picked up the merchandise in Acapulco and delivered it to the cities where the merchants resided. When entering the city of Mexico, each caravan owner had to register the cargo. Raw silk, silk and cotton clothes were more common among the products purchased by the merchants. Ceramics and storage jars were also goods frequently brought into Mexico City. It is difficult to know how much benefit these Asian products brought to New Spain, although fragile items, such as furniture and ceramics, were always among the goods transported to Mexico by land and which were probably sold at higher prices. The route used to transport goods from Acapulco to Mexico was known as *Camino de la China* (The Chinese Route). The actual 16th-century route is not known and Francisco Carletti's famous journal account of his travels by land from Acapulco to Mexico remains short or details and obscure. The chronicle of Sebastian Cubero, who travelled from Acapulco to Veracruz in 1670, refers to a route that passed by Papagayo, the Papagayo River, Tixla, Chilapa, Atlixco, Puebla and Veracruz.[31] In 1697, Francisco Gemelli Careri passed by the Papagayo River, Cañahuatal, Dos Caminos, Acahuizotla, Mazatlán, Las

---

[26] Sofia Sanabrais, *The Biombo or Folding Screen in Colonial Mexico, Asia & Spanish America, Trans-Pacific Artistic & Cultural Exchange, 1500-1850*, Oklahoma University Press, Denver, 2009, pp. 69-106.

[27] Sonia Ocaña Ruiz, 'Enconchado Frames: The Use of Japanese Ornamental Models in New Spanish Painting', *Asia & Spanish America, Trans-Pacific Artistic & Cultural Exchange, 1500-1850*, Oklahoma University Press, Denver, 2009, pp. 129-150.

[28] Etsuko Miyata Rodriguez, 'The Early Manila Galleon Trade: Merchants' Networks and Markets in 16th- and 17th- Century Mexico, *Asia & Spanish America, Trans-Pacific Artistic & Cultural Exchange, 1500-1850*, Oklahoma University Press, Denver, 2009, p. 43.

[29] María Ramón Serrera, *El Camino de México a Acapulco*, El Galeón de Manila, Aldeasa, Ministerio de Educación, Cultura y Deporte, 2000, pp. 41-42.

[30] Chinese ceramics found in Florida and Panama may be viewed online at the website of the Florida Museum of Natural History: https://www.flmnh.ufl.edu/histarch/gallery_types/open_search_proc.asp Luis A. Romero, 'La cerámica de importación de Santo Domingo, Antigua Guatemala', in J. P. Laporte, B. Arroyo y H. Mejía (eds.), *XX Simposio de Investigaciones Arqueológicas en Guatemala*, Museo Nacional de Arqueología y Etnología, Guatemala, 2006, pp. 1529-1545 (digital version). Mónica Pavía Pérez, *Arqueología Subacuática en Cuba. Reseña histórica*, Habana Patrimonial, Dirección de Patrimonial Cultural, Habana, 2011, online at: http://www.ohch.cu/articulos/arqueologia-subacuatica-en-cuba.-resena-historica/
Meiko Nagashima, 'Japanese Lacquers Exported to Spanish America and Spain', *Asia & Spanish America, Trans-Pacific Artistic & Cultural Exchange, 1500-1850*, Oklahoma University Press, Denver, 2009, pp. 107-118.

[31] Serrera, Ramón María, *El Camino de México a Acapulco*, *El Galeón de Manila* (catalogue), Aldeasa, Ministerio de Educación, Cultura y Deporte, 2000, pp. 41-42.

Petaquillas, Chilpancingo, Zumpango, Rio de las Balsas, El Nopalillo, Pueblo Nuevo, Amacuzac, Ahuacuotzingo, Alpuyeca, Cuernavaca, and Tlatenango. This route referred to by Francisco Gemelli Careri did not change until the cessation of the Manila galleon trade in the 19th century, and it is still used as federal road.

Asian products brought by the Manila galleon trade were acquired mostly by merchants of New Spain, through 'agents' residing in Manila. These agents were normally family members of the merchants or business partners. They received orders from their offices in New Spain or Peru and purchased goods from the *sangleyes* (Chinese) or Portuguese in Manila, or whoever came with goods to Manila. On the other hand, merchants in Mexico, La Puebla, Oaxaca, Veracruz, and Lima also had relationships with the merchants in Spain and sent orders from merchants in Seville or other cities to Manila via their agents. In other words, some of merchants in Spain bought Asian goods via merchants in Mexico or other large cities. These merchants in New Spain formed an association referred to as *consulado* (consulate), taking its model from the Basque and Catalonian merchants' association. The *consulado* facilitated commercial activities in New Spain by their own budgets. It also managed a cartel among the merchants to fix the price of imported goods and independently functioned to benefit their trading activities. Most of the merchants of New Spain who controlled these imports of Asian goods to Spain were from either Mexico or Veracruz.[32]

Veracruz to Seville

Some of the Asian merchandise brought to New Spain was then in turn transported by land to Veracruz, and from there shipped to Seville, which was one of the largest and most important ports in Europe in the 16th and 17th centuries. It was a port where all the goods from America, especially silver, could enter with other goods, such as leathers and dye materials. Although Asian goods were never the major products traded in Seville, they did come through the port and were consumed in Spain. Silk was traded on a fairly regular basis, but other luxury goods, such as ceramics, were very small in number but were traded as personal gifts. Most of these items were probably consumed in Seville, the city where wealth concentrated within Spain. The distribution of Asian goods within the Spanish territory hardly spread towards the interior regions. The only part of the country which always traded in certain Asian goods, especially Chinese porcelain, was the coastal Galician province located in the northwest of Spain.[33] This phenomenon was the result of geographical factors, the region being adjacent to Portugal and traditionally maintaining active commercial relations by sea routes. It will later be discussed that the earliest Chinese pieces entered Spain via Lisbon and the Galician coast, and that these were possibly moved further north to Amsterdam, which was destined to become the largest consumer market for Chinese ceramics in 17th-century Europe.

Spanish markets for Asian products

How Asian products were circulated and accepted in the Spanish society of the 16th and 17th centuries is still not very clear as there are so few Asian products found as heirlooms or excavated materials, although they did have some influence directly or indirectly on Spanish decorative arts, such as ceramic productions. Many of the Spanish ceramics of the 17th and 18th centuries copied the Chinese blue-and-white motifs of the late 16th and the early 17th centuries. Portugal in the 16th century was the largest importer of Chinese ceramics in Europe and traded large quantities from the beginning of the Asian expansion in the early 16th century.[34] The Netherlands took over from Portugal in the 17th century after the establishment of the Dutch East India Company and became the largest importer into Europe. Compared to these countries, which were competitors to Spanish influence in Asian waters, Spain imported very few Chinese ceramics. Later, England, and Germany too, began to import Chinese ceramics, and in the 18th century almost all the western European countries created certain levels of demand for Chinese ceramics and this boom became a general social phenomenon. However, in case of Spain, imports of Chinese ceramics were fairly modest from the 16th century, when the Manila galleon trade began, and was able to import Asian goods for New Spain. Although the 18th century shows a different scenario, with more imports of Chinese polychrome ceramics, it must be said that the quantities were still limited compared to other European countries.[35]

Each component of the Asian trade, trans-Pacific trade, American trade, and Atlantic trade was organized and connected by the Chinese, Malays, Japanese, Portuguese, Spaniards, Mexican and Dutch merchants. Within this large scheme, Chinese ceramics are one of the commodities that will enable us to understand the actual nature of the trade route and the distribution of Asian goods to Europe by the very fact that ceramics survive whereas many so other materials (wood, paper, textiles, etc.) eventually disappear. Studying the spatial distribution of ceramic materials from various excavation sites demonstrates how Chinese ceramics were traded in Mexico and Spain together with other Asian products. The presence (or absence) of material

---

[32] See Appendix.
[33] Etsuko Miyata, 'Chinese ceramics from Spain: their significance in the 16th and the 17th century Atlantic coastal trade', *52 Congreso Internacional de Americanistas 2009*, online distribution.
[34] Jean Paul Desroches, 'Oriental Ceramics and Porcelains', *Nossa Senhora dos Mártires*, The Last Voyage, Verbo, Lisbon, 1998, pp. 233-234.
[35] See Appendix. This issue will also be discussed in Chapter IV in detail.

from specific periods shows the rise and fall of trade activity as influenced by the cultural trends of the time and the economic conditions. The general tendency of the Chinese ceramics trade in Asia, America and Europe shows that the end of 16th century and into the early 17th marks the zenith of such trade, and a great quantity of ceramics of the Ming (明) period were exported over almost all the world. The mid-17th century was a time when many societies experienced some form of economic crisis and, especially, it is believed that Jingdezhen (景德鎮) production ceased by the time of the civil war that occurred at the beginning of the Qing dynasty (清朝). Consequently very few pieces of this period are present in the excavation sites in Mexico, Portugal, the Netherlands and Spain. The types of ceramics also show a change over the course of time, adapting to the demands of each market. When the ceramic trade with European countries began in the early 16th century, first with Portugal, the forms, which were completely based on Asian prototypes, began to evolve to suit the cultures of Europe and America (i.e. flat plates, coffee cups and saucers). For example, from the third quarter of the 17th century, tall, thin cups began to be produced and can be found in various excavation sites in Mexico and Europe. These were probably used to drink coffee or chocolate, and in the 18th century, large quantities of cups and saucers were produced and exported to the European market. These forms can hardly ever be found in excavations in Asia, except for some port-cities where Chinese ceramics were loaded and exported by Dutch merchants. Henceforth, the presence of a specific shape in one culture, and its absence in another, indicate the difference in eating/drinking habits and cultures between countries. The variety of types and shapes of Chinese ceramics increased greatly from the third quarter of the 16th century in order to suit the needs of export markets; it continued to increase until the 18th century, when polychrome ceramics became popular in European societies instead of the earlier blue-and-white style. This probably reflected the tastes of the European market, where traditionally colourful ceramics were more favoured, as can be seen on Majolica ceramics in Italy and Spain, whereas in Asia, blue-and-white wares continued to be the major preference. Ceramics as archaeological material inform about material culture and its change over time, strongly related to the economy and trading activities in each society or region.

Chapter I

The Arrival of the Portuguese and Spanish in Asian Waters

1. The expansion of Iberian power towards Asia

This first chapter will explain the background and some elements that made the Manila galleon trade develop and flourish from the 16th century. The arrival of the Spanish in the Philippines, the founding of Manila and maintaining Spanish interests in Manila were considerable challenges for Spain – with the country's limited knowledge of the islands and region. The Spanish on their own were never able to realize this growth in trade between Manila and Acapulco, which exchanged various products from China and other countries from Southeast Asia for silver. As mentioned previously, it was due to the historical composition of this period that Asia was able to enter a phase of commercial expansion that sparked the galleon trade. Many countries in the majority of Asian regions were able to construct complex networks: from India towards Southeast Asia and the South China Sea to Japan, and it is likely that the Portuguese participated in this trade using the existing network and taking control of Goa and Malacca. Tomé Pires visited all the important port polities from the Middle East to eastern Asia and commented on all the relations between port-cities.[36] Trade in Southeast Asia (and contraband trade along the southern China coast) carried out by the Chinese and the Portuguese brought the Portuguese not only wealth but probably also important knowledge of mercantile networks and connections with merchants in yet unknown regions. When Macao was officially founded, the Portuguese were trading with the Chinese and starting to exchange silver for silks in Japan as well. When Miguel López de Legazpi arrived in Cebu, he had little idea how to administrate these islands, with very few products of value to sell. However, he came to the Philippines at an opportune time, when Asia as a whole was active and receptive to trade. The Portuguese were already there, having spent more than 50 years in this region and integrating Manila as one of the intersecting points in Asian trade using American silver. It may be claimed that without the existence of the Portuguese and the base of Macao in Asia, the Manila galleon trade would never have been able to start and develop as quickly as it actually did. These elements to do with the founding of the new and extensive trans-Pacific trade network are important factors that will be discussed further in the next chapter.

What was the prime mover that drove these two countries towards Asia in the 16th century? According to Charles Boxer, '[If] it was the search for the Christians and spices which brought the Portuguese to Asia, it can be said that Christians and silver were the twin lodestars of their annual voyages to Japan.'[37] However, spice, silk, porcelain and other luxury goods have always had always been greatly desired and appreciated in Europe and Arabian countries as early as the Roman period, and, thus, it could have been the Italians, Arabs or any other European power who had the motive and resources to come to Asia for trade in a significant way.

A Roman coin from the 1st century was excavated in the Cambodian kingdom of O'keo, signifying that there was trade between the two regions. Large numbers of Chinese ceramics from the 10th to the 15th centuries have been excavated in many Arabian ports, such as Hormuz and Fustat (Old Cairo).[38] Even from the Muslim fortress at Almeria, which had a trade link with North Africa, Chinese white-ware of around the 11th century can be found.[39] It is a well-known story that Marco Polo brought back 13th¹-century Jingdezhen qingbai ware (青白磁) to Italy. Such incidents are just a fraction of the evidence that Asia and Europe had some form of trading relationship from the earlier periods, probably carried out first by Arab and then Italian merchants.

The Iberian powers, Portugal and Spain, were both involved in the *Reconquista*, until 1249 in the Algarve in Portugal and 1492 with the fall of Granada in Spain. After the establishment of the two kingdoms, Portugal gained independence from Castile in 1411 and both began expansion overseas. Portugal first began its campaign by attacking Ceuta in 1415 and further exploring the African coast, under the banner of evangelization, leaving *padrões*, stones with engraved Portuguese escudos, to claim possession of the territory. The conquest of Madeira (1420) and the Azores (1427), under the sponsorship of the Infante Don Henry the Navigator, further encouraged Portugal's political and economic strategies to advance southwards. Portugal brought slaves from Africa and transported them to Madeira to work on newly founded sugar plantations; sugar was distributed to Europe by Genoese and Flemish merchants. Antwerp became an important destination for sugar transported from Lisbon.

---

[36] Tomé Pires (Shigeru Ikuta translation), *Tohoshokokuki*, Daikokaisousho, 1973.

[37] Charles. R. Boxer, *The Great Ships from Amacan, Annales of Macao and the Old Japan Trade, 1550-1640*, Centro de Estudos Historicos Ultramarinos, Lisboa, 1959, pp. 1-10.

[38] Tsuguo Mikami, *Boueki Tojishi Kenkyu*, Vol. 3, Chuokouronbijutsushuppan, Tokyo, 1988, pp. 9-15.

[39] David Whitehouse, 'Chinese Porcelain in Medieval Europe', *Bollettino d'Arte*, 1966, pp. 63-73.

Spain was finally united with the marriage of the Catholic monarchs in 1474. One of their strategies to merge the diverse provinces was the undertaking of the nororious Inquisition, with the establishment of the *Santo Oficio* (Holy Office) in the 1480s, mainly in Andalucia, and concluding with the Jewish expulsion in 1492. The Inquisition forced many Jewish merchants to move north, although the main destination was adjacent Portugal. Some of the Jews who converted to the Catholic faith, generally called *conversos*, chose to integrate with Spanish society, engaging in banking, commerce, medicine, with some rising socially to become royal officers and religious elite. Major *conversos* merchant families, especially those residing in Seville, which was one of the most prosperous cities in Europe at the time, became linked through marriage and spread their commercial networks extensively, especially after the discovery and conquest of America, by sending family members abroad as agents.[40]

Spain conquered the Canarias in 1478, founding sugar plantations and bringing great wealth to the region. Slave labour was key to these enterprises, and Seville and Lisbon in the 15th and 16th centuries were the two largest cities in Europe engaged in the slave trade. Seville traditionally had a large market of Muslim, African, and 'Lolo' (mixed-race) slaves. Significant numbers of slaves to the Canarian sugar plantations were brought by the Portuguese from Guinea and Angola, via Lisbon and then on to Seville to be transported to the Canarias.[41]

Portuguese influence reached Southeast Asia in 1509 (Malacca), and Spain in the Philippines (Cebu) in 1521, which marked the beginning of their participation in the Asian trade.

It may be useful to consider similarities between Lisbon and Seville when we think of the main drivers of Portuguese and Spanish overseas expansion. First, when the Atlantic trade route began to play an important role in European trade in the 15th century, the two cities were both located at important locations where ships stopped over and consequrntly they were already important and wealthy port-cities with fairly large populations. Second, as commercial cities, both were centres of merchant migration from Geneva and Flanders, with traders active and eager to invest in new overseas business. Third, the population of *conversos* was also an important factor for these two cities and countries since many were already investors in overseas expansion as merchants, negotiators and interpreters in these little known places.

2. Trade in the Asian region before the arrival of the Portuguese and Spanish and their first contacts

Historically, many countries in Asia around the Indian Ocean and Chinese waters were closely involved in sea trade owing to their geographic nature, being islands, peninsulas and coasts. Trade activities during 'the Age of Commerce' in Southeast Asia and neighbouring regions have been studied by numerous scholars and further new theories and discoveries have sprung from recent research of shipwrecks and the traded ceramics they contain.

When the Portuguese arrived at Goa, Malacca was already an important commercial port where many products from Southeast Asia, China and Japan were traded. Pegu, Pasai, Tuban, Gresik, Ayutthaya, Champa, Cochin, Brunei, Zhangzhou, Guangdong, Sakai and Hakata were all active port-cities at the time. China exported raw silk and Jingdezhen wares, which were both highly valued all across Asia. Annam also exported ceramics of good quality that were prized in Ryukyu and Japan. The Ryukyu kingdom was another port polity that grew in influence during this pre-European period from its three-way trade with China and Japan. Ryukyu merchants it seems also played a similar role to Portuguese merchant during the 16th and 17th centuries, acting as middlemen during the Ming ban on commercial activity along the coastal area, following the losses inflicted by Wako pirates, and no official trade was permitted between Japan and China. The Ryukyu kingdom was divided into three political powers and was finally unified in 1462. The king sent tribute twice a year to Fujian, bringing horses and sulfur, and in return the official merchants were given Chinese products – silks, ceramics, silver and copper coins. The Ryukyu tributes sometimes included brazilwood (pernambuco), peppers, ivories, incense and cloves, acquired from Southeast Asia. They brought back these Chinese goods, of which silk was the most valued, and so undertook trade in Japan and Southeast Asian countries. Ryukyu merchants traded frequently in Malacca, Ayutthaya, Champa, Annam, and Patani.

Ayutthaya exported rice, forest products and ceramics, which began being mass produced in Sawankhalok during the 15th century. Other produce, such as liquors, were exported in container jars produced in Sawankhalok and, later, Sinburi. Many of these container jars are found in shipwrecks and especially in the Ryukyu Islands, which gained wealth from trade with Japan and China. Japan exported silver, which began to develop its mining technology in *Iwami* (石見) (Shimane Prefecture 島根県), and weapons. Additionally, short-distance trade within the southeast region was undertaken in smaller boats, transporting rice, vegetables, dried fish, livestock, palm wine and salt for urban consumption.[42]

---

[40] Luisa Schell Hoberman, *Mexico's Merchant Elite, 1590-1660: Silver, State, and Society*, Duke University Press, Durham and London, 1991, pp. 43-44.
[41] Hirotaka Tateishi (ed.), *Daikoukai no Jidai-Spain to Shintairiku*, Doubunkan, Tokyo, 1998, pp. 3-38.

[42] Anthony Reid, *Southeast Asia in the Age of Commerce 1450-1680*,

The Portuguese were quick to see the potential wealth to be had from the Spice Islands, based on Malacca, as well as numerous other products from the wider Asian region.

When the Portuguese first arrived in Malacca, an anonymous cartographer remarked '[In] this city, there is every kind of merchandise that comes from Calicut, to wit, cloves and benzoin and lignaloes and sandalwood, styrax and rhubarb and ivory and most valuable precious stones and musk and fine porcelain and much other merchandise; all, for the most part, come from the Land of Chin'.[43] This is the first informative report on China, a vague account, but somehow giving hope to the Portuguese that there was a land of riches out there somewhere to be exploited. In 1511 Afonso de Albuquerque conquered Malacca, and before he left his mission he sent his pilots to attempt to establish friendly relations with the neighbouring states and distribution centres for various items from Siam to the Moluccas, including Sumatra, Java, Timor Borneo and Luzon.[44]

Products distributed from these areas included camphor, benzoin, mace, musk, cloves, nutmeg, lacquer, silk, porcelain and many others. Thus the Portuguese subsequently began to widen its trading influence in every direction from Malacca, embracing practically all the Asian ports. Even before the arrival of the Portuguese, Malacca was already an important trade centre, and, according to Pires, merchants called from Cairo, Mecca, Moros, Abisian, Kiluwa, Melindi Ormuz, Persia, Turkey, Tulkman, Armenia, Gujarati, Goa, Daken, Malabar, Kelin, Orissa, Ceilon, Bengala, Arakan, Pegu, Siam, Kedah, Malayo, Pahan, Patani, Cambodia, Champa, Cochin, China, Lekeu, Burney, Luzon, Tanjungpura, Bunka, Linga, Sunda, Palembang, Jambi, Tunkal, Andalgeri, Kapo, Cambal, Minankabau, Siak, Lubato, Alkato, Maldives, and certain other locations as well.[45]

Pires focuses on those products originating from China: large quantities of white raw silk, coloured raw silk, damasks, enrolados,[46] taffeta, and many other varieties of silk textiles. Other items included pearls of various sizes,[47] musks of good quality, camphor, sulpher, bronze, iron, saltpeter, alum, rhubarb, jars, iron pots, basins, bowls, fans, needles, bracelets, and countless forms of porcelain. These goods come to Malacca as tribute, as Pires notes, with some originating from China, i.e. white raw silk (from Zhangzhou or Quanzhou) and porcelain, as well as coloured raw silk from Cochin.

The same author also relates that there are many other small islands around, and from there come slaves and stores of rice to be traded in Malacca. Merchants from Cairo, Mecca, Aden, Persia, Ormuz, Moro, Turkey and Armenia travelled to the kingdom of Gujarati with many products, and some undertake trade in this kingdom in return for cotton garments and from there sail on further to Malacca together.[48] Products from Cairo included weapons, woven textiles, corals, bronze, silver and mercury, iron and crystal and crystal wares decorated with gold. From Mecca came large quantities of opium, rosewater and brazilwood; from Aden there was opium, raisins, liquors, indigo, rosewater, silver, mother-of-pearl and dyes. All these commodities were destined for the markets of Gujarati. In Malacca these merchants brought back mace, nutmeg, sandalwood, mother-of-pearl, small quantities of porcelain, musk, benzoin, gold, raw silk, tin and silk.

After the Portuguese took Malacca, they sent ships to Java, Banda, Pasai, Paleacate, Timor, Martaban and many other locations. Within the town of Malacca, besides Malay merchants from Keling, other dealers from Java, Persia, Bengala, Pasai, Pahan and China were engaged in trade.[49] Many Malaccan merchants brought their own ships, and others traveled in junks or petacas owned by other idividuals, and from there sailed to places such as Sunda, Tajungpura, Pasai, Kedah, Siam, Pegu, Bengala, Paleacate and China. Trade with Sunda was particularly beneficial as merchants were able to bring back slaves and black pepper.

Joao de Barros refers to several types of metals, such as gold and tin from Sumatra, silver from Siam and bronze from China. All these were traded in Malacca in return for spice, medicinal plants, incense and silk.[50] Some interesting descriptions of how the products were packed and transported aboard can be found in his account. Mercury packed in Asia by Muslims had two types of container: coco shells and bamboo sealed with beeswax. Textiles were rolled tight and packed in hemp bags and closed with nails. These were then coated in tar and bailed. When the goods are transported to the port and stored, the duty officers weighed each package and registered them.[51] This stystem of cargo recording was probably also employed by the Spanish in the 16th and 17th centuries.

The diplomatic relations between China and the Portuguese after their arrival and the context of Asia during the period of extensive maritime trade has been well studied by Yoshitomo Okamoto.

Volume Two: Expansion and Crisis, Yale University Press, New Haven and London, 1993, p.67.
[43] Rui Manuel Loureiro, *Pelos Mares da China*, CTT Correios, Lisbon, 1999, pp. 24-28.
[44] Rui Manuel Loureiro, *op. cit*, pp. 28-29.
[45] Tomé Pires (translated by Shigeru Ikuta), *Tohoshokokuki*, Daikokaisousho, 1973, p. 453.
[46] *Enrolado* is usually a thin textile woven in India, but here it is of Chinese origin.
[47] Tomé Pires, *op. cit*, p. 244.

[48] ToméPires, *op. cit*, p. 458.
[49] Tomé Pires, *op. cit*, p. 488-489.
[50] João de Baros (Shigeru Ikuta translation), *Asiashi*, Daikoukaisousho, Tokyo, 1981, pp. 22-23.
[51] João de Baros, *op. cit*, p. 447, Cartas, I, 167-71.

Two years after the conquest of Malacca, the Portuguese sent ships to China, and in 1514 they sent more. In 1515 Jorge Alvarez went to China, and in 1516 eight ships were sent under the command of Fernão Peres de Andrade.[52] This increased level of contact with China on the Portuguese side indicates the significant interest in China – the largest and most powerful empire in the East. Andrade's ship included the first official embassy lead by Tomé Pires. The party first landed in Guangdong and waited for permission from the Chinese authorities to head north to Beijing. In 1520 Pires and his group finally set off for the capital, although they had to wait for the emperor to return from his travels the following year. During this time, the Chinese authorities were informed of disorderly trading activities carried out by the Portuguese along the south China coast, as well as a complaint from the expelled sultan of Malacca about his treatment at the hands of the invaders.[53] Emperor Zhengde (正德帝) died in that same year and Tomé Pires and his men were ordered to return to Guangdong without receiving an audience. When Pires returned to Guangdong he faced hostility from the Chinese who claimed that the Portuguese had plundered and kidnapped locals to sell as slaves. Owing to this incident the Chinese authorities banned all trade with the Portuguese (c. 1521) and Pires and his men were imprisoned, having all their gifts intended for the emperor confiscated.[54] This was the complete closure of Guangdong by the Chinese to the first European visitors.

The founding of Malacca permitted the Portuguese to try and access Asian ports and their interest in trading with China was central. Some early pieces of Chinese porcelain excavated in Lisbon are dated late 15th/early 16th century. The Portuguese, therefore, may already have begun trading somewhere near Malacca even before 1511, or may have acquired them in Malacca from merchants from other countries. After the ban on trade with China in Guangdong it took several decades to open official relationships with China. With the establishment of Macao in 1557 the Portuguese could begin what was to become a significant smuggling operation in terms of Chinese products. According to the study of James K. Chin, after the prohibition Portuguese activities along the China coast were focussed on Zhejiang,[55] which is also referred to by Donald F. Lach, although the latter clearly indicates *Lianpo* (双嶼島), off Zhejiang, as being the hub of clandestine trade, where Japanese and Chinese private merchants met.[56] This relationship between the Japanese and Chinese in *Lianpo* (双嶼島) later brought the Portuguese to Japan (c. 1543) and initiated the joint silk and silver trade which will be discussed in more detail later in this chapter. This role as middlemen between China and Japan, when official trade was banned between the two countries, firmly consolidated the Portuguese position and they were finally given permission to settle in Macao. The lack of any detailed documentation during this period of early clandestine trade activity can be partly filled by the existence of items of Chinese porcelain. More than 15 sherds of Jingdezhen blue-and-white ware (early 16th century) have been excavated from the site at Alfama site (see Chapter 4), and several fine and complete pieces, with particular escudos (c. 1530/40), can be seen in museums in Portugal (Photograph 1).

These finds are archaeological and historical evidence of this peculiar period when the Portuguese were developing a way to trade with Chinese merchants somewhere on the southern coast of China. Fernão

Photograph 1: Chinese ceramic sherds excavated from Alfama, Lisbon.(©Museu da Cidade, Lisboa)

---

[52] Yoshitomo Okamoto, *Jurokuseiki No Nichiou Koutsushi no Kenkyu*, Hara Shobo, Tokyo, 1974, p.99.
[53] Donald F. Lach, *Asia in the Making of Europe, Volume I, The Century of Discovery*, University of Chicago Press, Chicago, 1994, pp. 734-737.
[54] Donald F. Lach, *op. cit*, pp. 734-737.

[55] James K. Chin, 'The Portuguese on the Zhejiang and Fujian Coast Prior to 1550 as seen from Contemporary Chinese Private Records', *Macau During the Ming Dynasty*, Luís Filipe Barreto (ed.), Centro Científico e Cultural de Macau, I. P, Lisbon, 2009, p.120.
[56] Donald F. Lach, *op. cit*, p.737.

Mendes Pinto in his Peregrinação (printed in 1614) refers to the Portuguese expeditions undertaken from 1534 to 1542 along the Chinese coast.[57] Fernão Lopez de Castanheda also mentions in his account that there were more than 50 ports that were better than Guangdong.[58] Although the account of Fernão Mendes Pintos is frequently questioned, its credibility and his reporting on the expeditions at this time, with all the historical background in terms of its interest in trade with China, cannot be completely denied when we can see such evidence of excavated and heirloom pieces. One of the interesting pieces is shown in Photograph 2, a delicately formed ewer featuring the escudo of King Manuel I, but upside down. Another example in the same illustration is a Jingdezhen blue-and-white piece with the inscription of 'Jorge Alvarez' and '1552', also upside down. These were especially ordered by the Portuguese from Chinese merchants who were the agents of the Jingdezhen factories. The incorrectly designed inscriptions and escudos indicate that the Jingdezhen painters were not used to special orders, especially with Western lettering, and these pieces may be among the earliest ordered for the European market during the years of unofficial trade in southern China.

During this period of illicit trading the Portuguese arrived in Japan and were said to have introduced the first firearm there. This historical episode is mentioned in various accounts of the time in Japanese, Portuguese and Spanish. The actual year of their arrival and contact with the Japanese is unclear since all the accounts refer to different years,[59] although between 1541 and 1543 is the generally accepted timeframe. The ship with Chinese merchants and three Portuguese is said to have been wrecked and washed ashore on the island of *Tanegashima* (種子島). Some scholars argue that the ship was owned by the famous Chinese pirate *Wangzhi* (王直) on their way to *Lianpo* (双嶼島) from Ayutthaya.[60] It is probable that these Portuguese traders came to Japan in a Chinese merchant junk since the Chinese already knew the route to Kyushu, and it is likely that they did not arrive in Japan by chance, but rather the Portuguese intended to come to Japan under the protection of Chinese merchants in the hope of starting a trading relationship. When we consider the timing of their arrival and the products (firearms) they introduced to Japan, it is much more logical to say that their journey to Japan was planned from the beginning with the aim of developing trade.

At that time Japan was a society of warring states, which meant that all the feudal samurais required weaponry. Firearms introduced by the Portuguese were much

PHOTOGRAPH 2: JINGDEZHEN BLUE-AND-WHITE EWER DATED TO THE EARLY 16TH CENTURY.(©FUNDAÇÃO ALAMEIDA)

prized. It might not be too hypothetical to say that the Portuguese were already informed of the Japanese political situation and offered something of great interest to the wealthy samurais in order to find a way to begin trading. The Portuguese were also dedicated at the state level to missionary and therefore Jesuits were allowed to build churches and seminaries in some regions, for example in today's province of *Yamaguchi* (山口県), and providing weapons in exchange. The Portuguese acted as middlemen in trading between China and Japan as official trade was prohibited by China following all the problems caused by the *wakos* (倭寇) plundering along the southern coasts of China and Korea. *Wakos* first appear in Chinese documents from 1358, after attacking the eastern side of Yellow Sea and along the southern coast of the Korean peninsula. They were formed into small groups of two boats or so, with 20 to 30 men; each of these groups had no organized structure or relationship between one another.[61] These are generally referred to as 'early *wakos*', but in later years they became more organized pirate forces, mainly consisting of Japanese fighters but using Chinese or Korean guides as they marauded the shores of Korean and Chinese waters. Sometimes they would cooperate with Chinese pirates from the southern coast. These *wakos* were an obstacle to the maintainance of official Chinese–Japanese trade, and the Ming ban on commerce with Japan was

---

[57] Rui Manuel Loureiro, *op. cit*, p. 62.
[58] Rui Manuel Loureiro, *op. cit*, p. 62.
[59] Gakusho Nakajima, Portugal jin no Nihon hatsuraikou to Higasiasia kaiikikoueki, *The Shien* 142, Kyushu University, 2005, pp. 33-72.
[60] James K. Chin, *op. cit*, pp. 133-134. This has also been discussed in the Nakajima article above.

[61] Makoto Ueda, *Umi to Teikoku, Chugoku no Rekishi Ming Xing Jidai*, Kodansha, Tokyo, 2005, pp. 94-96.

announced in 1371. During this period the merchants of the Ryukyu kingdom benefited from their role as middlemen between Japan and China, selling Chinese silk and porcelain to Japan and bringing Japanese silver and other Southeast Asian goods to China. The Chinese economy was shifting to a silver monetary system from around the middle of the 15th century; tax began to be paid in silver and the government was in need of silver to administer the empire. In Japan, two Korean engineers were sent from *Hakata* (博多) (Fukuoka Province福岡県) to the *Iwami* silver mine (Shimane Province島根県) to introduce cupellation in 1533, and silver production increased rapidly. As Japan became the largest silver supplier in Asia, *Ouchi* (大内) began to export silver to the Ming through the *Kamiya* (紙屋) family, well known as one of the rich *Hakata* 博多) merchants.[62]

However, it might have been easier for the Japanese feudal lords to depend on the Portuguese rather than sending their own ships to China. In 1550 the Portuguese crown granted rights to travel to Japan to those members of the nobility who provided services to the monarch. This was referred to as *Capitão Mor* and the captain possessed total authority under royal patronage. The voyage was from Malacca to Japan, travelling for 35 to 45 days. The Portuguese brought spices and incense from Malacca and traded silk and porcelain and Guangdong. They would then sail back to Japan to acquire silver in exchange for Chinese goods. The return voyage would be the same, stopping over in some ports in southern China and then to Malacca. As mentioned, *Capitão Mor* was an exclusive authorization given by the monarch or the viceroy of Goa and produced large profits. However it is possible that other Portuguese private merchants were engaged in the China–Japan trade, since a captain in Malacca wrote to the Portuguese king in 1545 saying that '[More] than 200 Portuguese men'[63] were dispersed over the area to participate in trade outside the jurisdiction Portugal. It seems that the Portuguese diaspora in Asia was already formed by the first half of the 16th century. This owed much to the legal system of the Portuguese in Asia which established that when settling in Goa only married men (or women) were permitted to reside with official citizenship, and thus many soldiers who came to Goa and stayed single never had the possibility to become a citizen, and after coming to Asia many of them travelled independently and made their way as labourers, merchants or mercenaries.[64]

The Spanish arrived in Asia with Magellan's expedition and landed at Cebu in 1521. This famous expedition ended in tragedy with the death of the eponymous leader. It took several decades to discover a route between New Spain and the Philippines, crossing the Pacific, where they decided to settle and colonize. The Spanish in the Philippines were less active in trade within Asia, but once the Manila galleon trade began in 1565, and large quantities of silver were brought from New Spain, the Philippines became a region of great commercial interest to Chinese and Portuguese merchants.

It is difficult to grasp a detailed picture of how the Spanish began their settlement in Asia, and their relations with the Portuguese in the early phase of colonization, as there is a lack of written material. Around 1580, it is said that there were about 700 Spaniards in the town of Manila.[65] One of the earliest Spanish accounts of China appears in a document written in 1569[66] and based on information obtained from the Portuguese, '[When] the Portuguese were in this port, they understood how to negotiate and contract on the coast of China and Japan and how to negotiate. It is the Major Captain who brings this trade, which is partly granted by the Majesty in royal service. In some of the islets that we are newly discovering every day, we met two [men] from China with a boat and they had firearms. We have understood that these people have a large population and are well treated and the government well organized. There are big cities and thirteen of them are major ones. Their king is protected by three very strong walls with guards and he is at war with the Tartars…The names of these cities are Chincheo [Zhangzhou], Cantun [Guangdong], Hachina, Nimpoa [Ningpo], Onchiu [Wenzhou], Fuinan, Sisuan, Conce, Honan [Henan], Nanquin [Nanking], Paquin [Peking], Sachiu [Chuanzhou], Hucon, Latan, and Cencay…'

It is also noted in the same source (1567) that '[More] to the north from where we are, or almost northwest, there are large islands called Luzon and Vindoro [Mindoro]. Chinese and Japanese come there to trade every year and they bring silk textiles, bells, porcelain, iron, coloured cotton garments and other various goods. On their way back they load gold and wax. The people of these two islands are Muslims and they buy they products the Chinese and Japanese bring.'[67] This information is mentioned by the governor Miguel López de Legazpi in Cebu to the King. He also reports that in the Philippines there are not so many gold mines as expected and that the most abundant commodity is cinnamon. He also continues that at this time (1567) he still had little idea of what to do with these islands and was not sure of the benefits that any commerce with the Chinese and Japanese might bring. In this same document he writes that he was unable to be in touch with these merchants as he lacked soldiers and weapons enough to ensure

---

[62] Makoto Ueda, *Op. cit*. pp. 200-205.
[63] Rui Manuel Loureira, *Op. cit*, p. 74.
[64] Linschoten (translated by Seiichi Iwao), *Tohoshokokuki*, Iwanamishoten, 1973, Tokyo, p. 293.
[65] Antonio Morga, *Sucesos de las Islas Filipinas, Edicion critica y comentada y studio preliminary de Francisca Perujo*, Fondo de Cultura Economica, Mexico, 2007, p. 269.
[66] AGI Filipinas, 29, N.10, *Carta del factor sobre varios asuntos* Fols.31r-33v
[67] AGI Filipinas, 6, R.1, N.7, *Carta de Legazpi sobre falta de Socorro y descubrimientos*. Fols.1r-2v

their security. In letters of 1565,[68] 1567[69] and 1568[70] he repeatedly requests his need of munitions, ships (naos), money and manpower to secure their solitary base of Cebu, on the island of Visaya, where Legazpi had decided to stay until the order from the crown could be issued.

Curiously, however, as will be mentioned in the next chapter, Legazpi sent galleons in 1565, 1567 and 1568 from Panai to New Spain.[71] In 1570 he sent natives to the north and discovered Luzon. He refers again to the Chinese and Japanese who trade with the local Muslims and re-emphasizes the importance of having more people and weapons. He writes 'It is most important for the royal service to have ships [navios] here since the Spanish are not prepared to navigate in the ships of the natives.'[72] He also deployed the galleon *Espiritu Santo* in this year.[73] There are no previous studies on this very early stage of the Manila galleon trade and we have to depend on primary sources and the information sent by Miguel López de Legazpi. The argument here is that the governor of the Philippines on one hand insists that he is unable to negotiate with the Chinese or Japanese because of his lack of weapons and soldiers for five years, while on the other he appears actually to be sending trade ships to New Spain during this period. How should we interpret this contradiction that the Spanish in the Philippines were in a parlous state while at the same time sending merchandise to New Spain? The only interpretation of this is that the Spanish were trading with the Portuguese who came to Luzon with Chinese and other Southeast Asian products. It may be true that there was a lack of weapons and ships for the overall protection of a base so remote and isolated from its home country. However the documents that indicate the dispatch of galleons loaded with Asian products suggest that it was most probably the Portuguese who supplied these export goods to the Spanish.

Another observation that seem to substantiate this theory is that several sherds of blue-and-white plates (dated 1550-1570 at the latest) have been found at the excavated site of Templo Mayor in Mexico City (Photographs 3, 4, and 5). These are Jingdezhen ware of Occidental style with flat, wide rims, which began to be produced after the arrival of the Portuguese. Similar material has been recovered from the Portuguese wreck of the *Fort San Sebastian*, lost off Mozambique in mid-16th century.[74] These sherds are most probably from wares carried by the early Manila galleon trade as there are no similar examples in Spain. Either the Portuguese brought them via Malaca or they were acquired from the Chinese, indicating that trade existed before the establishment of Manila by New Spain (Photographs 3-5).

In any event, it does not seem coincidental that in 1571 Legazpi established Manila as the capital of the Philippines and was permitted to give an official license to the city in 1572.[75] That same year he received military support from New Spain and Manila began to be administrated as a capital. He was finally able to convince the king that colonizing the Philippine islands was feasible.

The establishment of Manila and its operation as a viceroyalty of New Spain was actually decided upon by the authorities of New Spain. However trade and political networking with other Asian and Southeast Asian countries and kingdoms were in many cases dependent upon the mundane activities of Portuguese and Chinese private merchants. In other words the vast Asian trade network from Goa to Japan was already fully understood, and based upon voyages from one port to another by the Portuguese and Chinese long before the establishment of Manila; the Spaniards had no other way to use their network and knowledge without sailing from Manila.

3. The establishment of Manila

In 1571, Miguel López de Legazpi founded Manila as the official capital of the Philippines and the base was given the title 'City of Manila'.[76] The reason Legazpi chose Manila as capital was because there was food in abundance – enough to supply all the Spanish population. The native Tagalogs made their living from the cultivation of rice, fishing, hunting and trade. They already had regular contacts with the Moluccas, Borneo, and Malacca, trading gold and food, and also dealt with the Chinese.[77] However, considering the previously mentioned account of Chinese and Japanese trade in Luzon, the most important reason for the founding of Manila as capital was this active regional trading nexus.

As the economy of the colonial Philippines was not based on the development of large-scale plantations or mining interests, even in its later years, the whole island was totally dependent on trade activity with New Spain and southeastern Asia, which was the major source of income (see Map 1).

---

[68] AGI Filipinas, 6, R.1, N.1, *Carta de Legazpi avisando su llegada y establecimiento*.Fols.1r-1v
[69] AGI Filipinas, 6, R.1, N.5, *Carta de Legazpi sobre descubrimientos realizados y armas*.Fol.1r
[70] AGI Filipinas, 6, R.1, N.8, *Carta de Legazpi pidiendo Socorro a Nueva España*.1 documento
[71] AGI Contaduria, 1196/1565-1576, *Caja de Filipinas, Cuentas de Real Hacienda*.
[72] AGI Filipinas, 6, R.1, N.12, *Carta de Legazpi sobre descubrimiento en el norte*. 1 documento
[73] AGI Contaduria 1197, *Caja de Filipinas, Cuentas de Real Hacienda*..Fols.126r-143v

[74] Mensun Bound, 'The Fort San Sebastian Wreck: A 16th-century Portuguese Wreck off the Island of Mozambique', Christie's, Amsterdam, 2004, pp. 2-11.
[75] AGI Patronato, 24, R.20, *Titulo a ciudad de Manila*. 1 documento
[76] AGI Patronato, 24, R.20, *Titulo a Ciudad a Manila*. 1 documento
[77] Nicholas Cushner, *Spain in the Philippines*, Ateneo Tuttle, 1971, p. 65.

Chapter I - The Arrival of the Portuguese and Spanish in Asian Waters

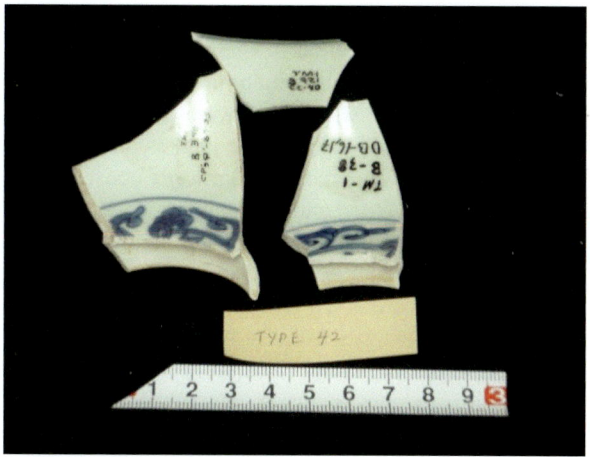

Photograph 3: Jingdezhen large blue-and-white basin dated to the middle of the 16th century (excavated from Templo Mayor, Mexico City, © INAH).

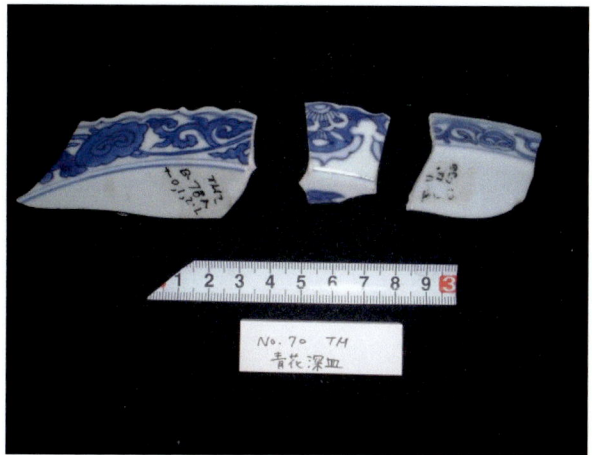
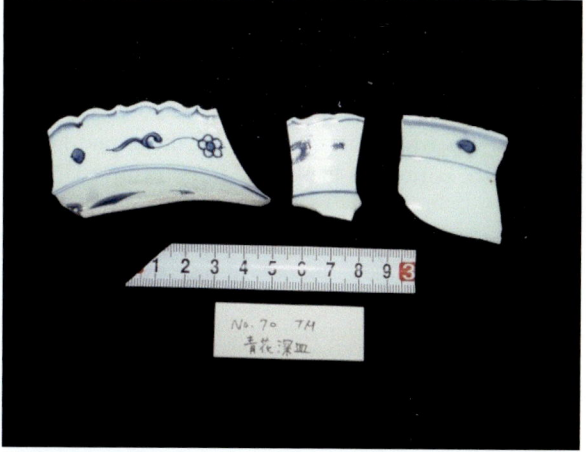

Photograph 4: Jingdezhen large blue-and-white bowl dated to middle of the 16th century (excavated from Templo Mayor, Mexico City, © INAH).

Photograph 5: Jingdezhen large blue-and-white bowl dated to middle of the 16th century (excavated from Templo Mayor, Mexico City, © INAH).

The detailed description of early Manila is provided in de Morga's account and other primary sources, the earliest of which is his document written in 1587. He writes that 'In the Philippines there are some Spanish residents and in the city of Manila, there are normally 700 and at times even 800 men. Although Your Majesty has had the foresight to offer each year help from New Spain, it has not always been carried out and when help comes there are very few people, and since they arrived here there have been no salaries or benefits. Some of them seem to be in great hunger and since the land is not healthy a large part of them die...All the houses and haciendas, including the fortifications, suffer from fire incidents and always in danger of burning as they are made of wood and canes and covered with nippa. It is not possible to construct roofs and strong and beautiful houses of brick and stone.'[78]

Despite the difficulties in ruling and administering the remote colony, after the establishment of the Manila galleon trade Manila became one of the hubs of Asian commerce, where products from China, Japan, Macao, Siam, Cambodia, Malacca, Macassar, Borneo and elsewhere were consolidated and redistributed. The materials involved included raw silk, porcelain from China, ivory, pearls, rubies, sapphires, cast iron wares, copper and rice. In Antonio de Morga's account, *Sucessos de las Islas Filipinas*, some products are mentioned in detail: 'In these islands there are mines and preparers of gold especially in the [islands of the] Pintados, the river of Botuan (Butuan), in Mindanao, and in Cebu, where they work in a mine called Taribon, with good gold… On some coasts of these islands there is mother-of-pearl, particularly in Calamianes…There are large sea turtles which the natives admire for their shells and which they sell as merchandise to the Chinese and Portuguese and other nations who come and look for them, and highly estimate the curiosities they make from them.'[79] He also mentions the various jars produced in the provinces of Manila, Pampanga and Ilocos 'that are made from very old brown clay, with seals, which [no one] knows where they were made nor when they were produced… The Japanese look for and value [these jars] as they produce herbs called cha. Kings and feudal lords conserve cha in jars which are valued and fetch high pricees.'[80] These vessels are the well-known 'Luzon Tsubo' (Luzon jars) that are still highly valued in Japan as part of the traditional tea ceremony. Recent academic research indicates that these jars were not actually produced in Luzon but most probably came from southern China, possibly Guangdong or Fujian, as simple containers, however the Japanese saw the aesthetic appeal of these vessels and they became high-status objects.[81]

Cinnamon from the island of Mindanao was much valued and other medicines and spices were gathered from Cavite, Compoc, Taya and Butuan. De Morga continues: 'Round peppers are found, although not in [large] quantities, but if seeded there would be as much as in the island of Cauchiu [Cochin] which is close to China. There are also *elefante* in the island in large quantities and also in the island of Joloc. There is ginger, tamarinds

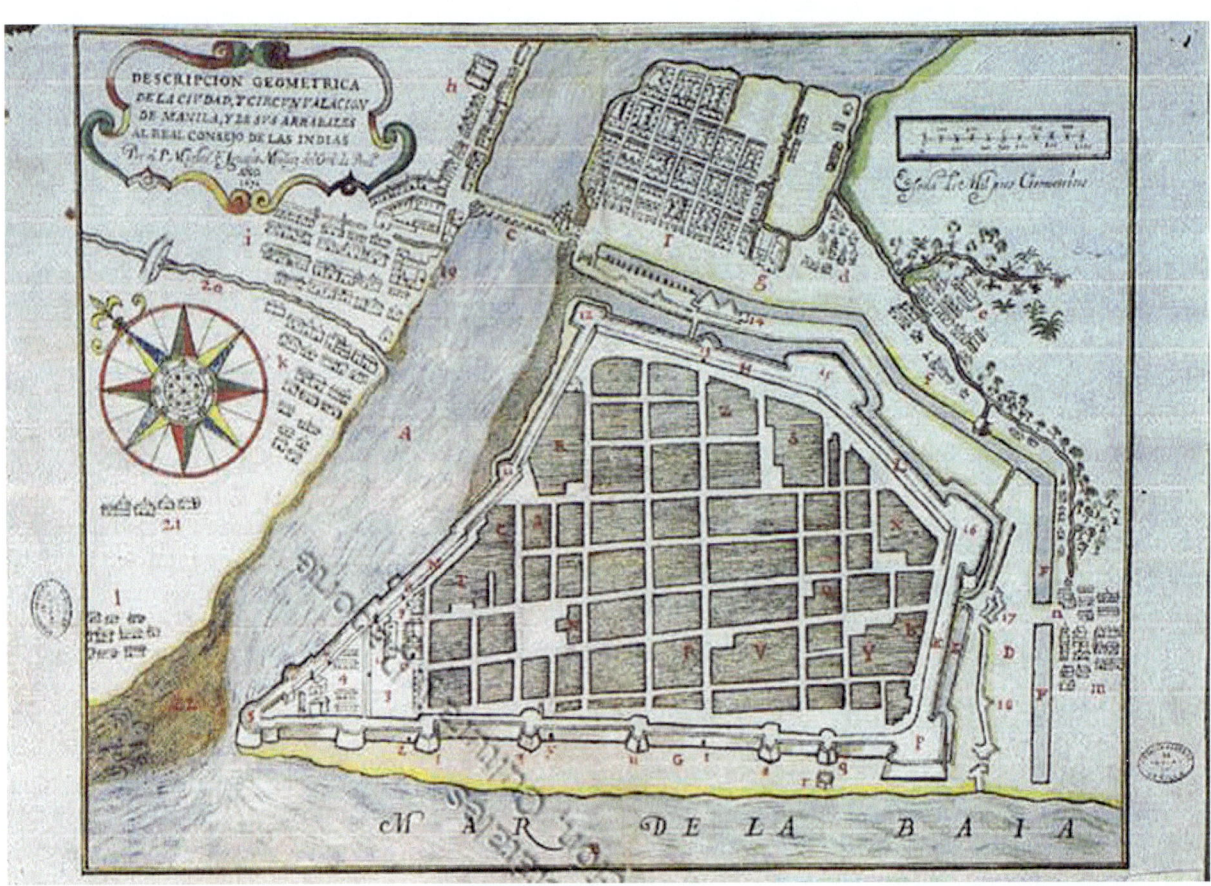

Map 1: Plan of Manila in 1671 ©AGI Filipinas, MP10.

---

[78] Antonio de Morga, *Sucesos de las Islas Filipinas, Edicion critica y comentada y studio preliminary de Francisca Perujo*, Fondo de Cultura Economica, Mexico, 2007, p. 269.
[79] Antonio de Morga, *op. cit*, p. 238.
[80] Antonio de Morga, *op. cit*, p. 239.
[81] Antonio de Morga, *op. cit*, pp. 239-240.

and other drugs.'[82] As we can see from this report, the Philippines only produced certain spices (e.g. ginger and cinnamon), cotton and gold. These commodities were not available in sufficient volumes to allow the region wide access to the dynamic opportunities of Asian commerce in this period.

4. Trade in Manila, Macao and other port-cities

As mentioned above, Macao was founded in 1557 and had contacts with the Spanish in the Philippines from the early phases of that nation's conquests in the region.

Portuguese interest in trade with Manila meant further trade in the Americas. The first petition was sent to the King in 1591, pointing out that trade between Manila and Macao been carried out without restriction and referring to the fact that 'Portuguese ships have always been to the Philippine Islands from Macao… In the past years, large ships from New Spain have come to the city of Macao and the Portuguese of Macao had monopolized trade with Japan and gained great wealth from it'.[83] In 1636 a petition was made to the king by Don Juan Grau Monfalcon, attorney general for the Philippines, that in the year before great harm was done to the residents of Manila by allowing 'the Portuguese of Macao to go and purchase merchandise in the Guangdong market [feria de Canton] in China and bring it to resell in the city of Manila… The Chinese came to sell merchandise in the said city and gave more favourable prices and credit until the return voyage without payment and sending the goods to New Spain and thus gaining much. All this has stopped with the arrival of the Portuguese who not only distrust but sell them with excessive prices, and if they are not paid, as they prefer, they send the products to Mexico, and they settle in Manila and keep the goods for one year or more, which the sangleyes did not do…'[84] In this document, not only does Juan Grau Monfalcon comment on Portuguese trade in the Philippines as damaging to the Manila merchants and residents, but also refers to the 1593 decree that all the Spanish are prohibited from going to the Guangdong market.[85] Nevertheless, Macao and Portuguese assistance were financial lifelines to the merchants of the Philippines, who could bring Chinese raw silk and other products to Manila. Although in 1616 the governor of the Philippines extracted a forced loan of 90,000 pesos from the Macao traders, the Portuguese still made a large profit and Manuel da Camara de Noronha reports that the galliots of Antonio Fialho Ferreira returned with a rich cargo of silver.[86] Ferreira was a resident in Macao before 1624 and was several times senior captain of the Manila trade expedition. He was also brother-in-law and business associate of Lopo Sarmento de Carvalho, who was probably a *converso*.[87]

The Portuguese in Macao acquired wealth in many ways from their involvement with Asian commerce. According to Boxer, the wealthy Cantonese merchants frequently shipped their goods to Manila through Portuguese agents.[88]

In 1636, a royal decree was issued confirming that 'Foreigners must not carry out business in these islands, including Portuguese, and that only the Chinese may trade freely; any Portuguese not holding a licence fom His Majesty will be expelled.'[89] Study of the primary sources from the late 16th century until the 1630s shows that the attitude of the Philippine government gradually changed in relation to trade between Manila and Macao. In the beginning, governor Luis Gómez Dasmariñas was keen on opening a free trade arrangement between Manila and Macao, and in fact there were ships sailing directly from New Spain to Macao, possibly to acquire Chinese products at first hand rather than dealing via Portuguese intermediaries. Subsequently the Spanish began to be less interested in going to Macao for fear of financial penalties resulting from shipwrecks. In the end the second order was issued so that free trade with Macao would be prohibited, although from time to time ships from Macao did come to Manila for that purpose. The products the Portuguese brought with them were not only raw silks and porcelains but various other goods such as wheat, rose-water, almonds, fans, cotton textiles, white cord, sweets, rice,[90] spices, slaves, rich quilts from Bengal, amber, ivory, jewels and precious stones.[91] Except for cereals, which for local consumption, the other commodities were luxury items for re-export to New Spain. From 1580 to 1644, 79 ships from Macao came to Manila, although it was not trade on a regular basis.[92] This was probably linked to the time when the Manila–Macao trade was most lucrative in relation to the Manila galleon trade.

The document further mentions that 'His Majesty reserves the right of *almojarifazgo* [customs duty] each

---

[82] AGI Filipinas, 84, N.3, *Relación del agustino Diego de Herrera para remedio de las Filipinas*. 1 documento
[83] AGI Filipinas, 27, N.30, *Carta del Cabildo secular sobre el gobernador das Mariñas*. Fols.166r-173v.
[84] AGI Filipinas, 41, N.16, *Petición de la ciudad de Manila sobre comercio de portugueses*. 5 documentos.
[85] AGI Filipinas, 41, N.16, *Petición de la ciudad de Manila sobre comercio de portugueses*. 5 documentos.
[86] Charles R. Boxer, *Macao 300 years ago*, Fundação Oriente, Macao, Lisboa, 1942, p. 134.
[87] Charles R. Boxer, *The Great Ship from Amacon, Annales of Macao and the Old Japan Trade, 1550-1640*, Centro de Estudios Históricos Ultramarinos, Lisboa, 1959, p. 102.
[88] Charles R. Boxer, *op. cit*, p. 12.
[89] AGI Filipinas, 82, N.1, *Copia de real cédula a la Audiencia sobre comercio de Portugueses*. 1 documento.
[90] AGI, Filipinas, N.70, *Autos sobre barcos San Pablo y Nuestra Señora de Piedad*. 2 documentos
[91] Charls R. Boxer, *op. cit*, p. 74.
[92] Manuel Ollé Rodríguez, 'Macau-Manila Interactions in Ming Dynasty', *Macau During the Ming Dinasty*, Luís Filipe Barreto (ed.), Centro Científico e Cultural de Macau, I. P. Lisbon, 2009, pp. 152-176.

Portuguese Intervention in the Manila Galleon Trade

year on any merchandise which the Chinese bring to this city from 80,000 to 100,000 pesos, and that those ships that come from Macao have to apply this *almojarifazgo* in any year above 20,000 pesos…',[93] and that 'The governors normally on occasion despatch from here some of Your Majesty's ships to the said cities of Malacca and Macao and these [Portuguese ships] bring back large quantities of slaves, and seeing that the fleet should pay tax, they come in Your Majesty's ships and prefer not to be asked to pay since they are not used to paying such taxes.'[94]

The Manila–Macao relationship was an ambivalent one in that, especially relating to the Manila galleon trade of the 16th and early 17th centuries, the arrival of Portuguese ships to Manila with luxury goods was essential, even if there was conflict between them. On the other hand, Chinese merchants who were also important for the Spanish Philippines are also frequently referred to as rivals to the Portuguese. However, the study of Atsuko Hirayama shows that *champanes* from China were mostly loaded with non-luxury items such as conserved foods, perishables, cotton cloths, horses, cows, lead, copper and iron, and were thus excluded from paying tax. Her study includes the find that several letters were sent from Manila to Spain saying that trade with Chinese merchants will never bring real benefit to Spain and should be substituted by commerce with the Portuguese. She concludes with three possibilities:

1. Since Manila was in permanent lack of munitions, provisions from the Ming Dynasty, which banned exports of weaponry, were not beneficial to the Spanish.
2. Manila–China trade might have been conducted by the Portuguese via Macao.
3. Goods were exported to Manila together with slaves by the Portuguese, and this trade was more attractive to the Spanish.

These are interesting possibilities and it is very probable that Macao and the Portuguese merchants were more active and important to the Manila galleon trade. Chinese merchants were, of course, important, but most of these might have been more interested in providing day-to-day supplies, coming not only to Manila but to Ilocos and Cagayan, with some operating as travelling vendors through the island.[95]

Other trade relations between Manila and various Asian countries were undertaken as well. From 1620 to 1640, 44 ships entered the port of Cavite – from Malacca,

---

[93] AGI Filipinas, 41, N.16, *Petición de Cabildo secular de Manila sobre comercio de portugueses*. 5 documentos.
[94] Filipinas, 29, N.57, *Carta de Francisco de las Misas sobre varios asuntos de Filipinas*. Fols.382r-404v
[95] Atsuko Hirayama, *Spain Teikoku to Chuka Teikoku no Geko, 16, 17 seikino Manila*, Hoseidaigakushuppan, Tokyo, 2012, pp. 281-284.

1657
Guangdong
Kingdom of China
Kingdom of China (a ship commanded by captain Andres de Zaretes)
Kingdom of Camboya (Cambodia)
120 picos of iron
200 picos of wheat
1500 bowls
1500 pieces of linen
780 white shawls from Japan
200 packets of paper brought by the Chinese
Kingdom of Cochin
500 white shawls from Japan
70 picos of iron from Japan
40 picos of wheat
Makassar (Nao San Antonio, owned by captain Juan Gomes de Paiba)
textiles
50 picos of peppers
Cochin China (a ship whose captain and owner is a Chinese merchant named Francisco Chosa)
White shawls from Cochin China
40 large plates
Batavia (Nuestra Senora del Sagrario, whose captain is Juan de Eruguesa)
200 picos of wheat
2 large boxes from Japan given by the Dutch
10 arrobas[1] of Castillan wine
6 frascos of Castillan oil
Kingdom of Sian (Siam)
500 picos of iron
1500 cotton shawls and other textiles
Kingdom of Sian (Siam) (a ship owned by Luis Hernándes)
textiles
Kingdom Makassar (a ship owned by one Magabilan, a Muslim of Makassar)
textiles
30 picos of pepper
Kingdom Makassar (a ship owned by one Charama, a Muslim of Makassar)
50 picos of pepper
textiles
A champan recently arrived from Japan, which has come from China; the ship owned by a Sangley named Siqua)
200 picos of wheat

1686
Kingdom of Golconda
Madrasta (coast of Coromandel)
Kingdom of Gujarati
Bengala
Kingdom of Sian (Siam)
sugar
plates

1687
Lianpo
Quanzhou

---
[1] 1 arroba = 11.5kg.

Maluku and India.[96] All the ships coming to Manila were

[96] Manuel Ollé Rodríguez, *op. cit*, pp. 152-176.

from countries under the rule or influence of Portugal. How Manila became one of the most important ports in Asia can be judged by the ships coming into Cavite during the second half of the 17th century.[97]

As can be seen from the above table, southern China, Siam, coastal India and Makassar were the regions which frequently sent ships to Manila. What is surprising is that, even until the end of the 17th century, the Portuguese were participating in some voyages as ship owners. Direct trade with Japan was completely banned by this period, although clearly indirect exchanges were common.

5. The Chinese in Manila

The Chinese in Manila were referred to as *sangleyes* (常來) in Fujianese dialect, which originally meant 'those who come frequently'. The first account of Chinese merchants in the Philippines appears in the document, previously mentioned, that the Chinese and Japanese were coming to Luzon to trade, although trade between China and the Philippines existed from the 10th and 11th centuries. Chinese ceramics from the Tang Dynasty are known from Butuan and many other blue-and-white wares have been found in the *Santa Ana* site, dated to the 14th/15th century Yuan period.[98] Shipwrecks have also been found loaded with Chinese porcelain and material of other provenance (Thailand, Burma, Vietnam), i.e. the Pandanan wreck site (Palawan Island, Philippines) dated to the late 15th century.

The vague accounts concerning the Chinese, as related by the Portuguese, might not have interested the Spanish as Legazpi did not get in touch with them right away. However, the Chinese began to come more frequently to the Philippines and some of them became baptized and resided in Manila and other places. The relationship between the Chinese and the Spanish was an ambivalent alliance/enemy one. The Spanish were always afraid of the large populations of Chinese who occasionally rioted against the authorities, but on the other hand it was the Chinese who supplied daily miscellaneous goods and labour to the Spaniards. In 1604 a document indicated that 13 Chinese ships came to Manila.[99] Regarding these Chinese visitors, most probably were only temporary, receiving licences from the chancellor in Manila. The document mentions that the Chinese ships came from Cebu and Mindanao.[100]

Captain	No. of *sangleyes* entering	No. of *sangleyes* returning
Jinuin	214	203
Pecan	247	178
Guansan	270	204
Binsan	182	140
Yanten	143	136
Sutian	290	229
Yonlin	135	211
Yachan	320	151
Siusan	80	119
Chenu	247	270
Guansun	178	185
Quibgou	150	57
Ontay y Guatian	230	87
Total	2686	2170

NUMBERS OF *SANGLEYES* ENTERING AND LEAVING MANILA IN 1604.

According to this written source there were more *sangleyes* staying in Manila than those leaving, and all of them, regardless of their profession or trade, were permitted to stay. At the end of the 16th century there were said to be around 6,000 or 7,000 *sangleyes* residents.[101]

The Chinese were engaged in trade and worked as tailors, shoemakers, embroiderers, silversmiths, tattooists, carpenters, wax chandlers, hatters,[102] fishermen, stonemasons, coal merchants, porters, bricklayers, jobbing labourers, etc.[103] Thus the fact that the Spanish in Manila could not flourish without the Chinese can be substantiated by this document of 1604.

In 1595 a law was enacted regarding the *sangleyes* to the effect that: 'The Chinese who come to Manila are to be limited to 50 men and the same number should leave the city… and that the Christians without faith who are normally here and come every year and live inside the city walls should be none.'[104] This is the first congregation of Chinese in Parián (San Gabriel), Binondo and Tondo, and they were transferred from *intramuros* to *extramuros* for the sake of the protection of the Spanish.

Chinese–Spanish relations were again tense in 1639, when large numbers of Chinese protested against Hurtado de Corcuera for forcing the Chinese to work in the rice plantations outside Manila. A subsequent threat to send tribute to Formosa by Koxinga in 1662 shook Manila and again a general feeling of mistrust against the Chinese or *sangleyes* arose amongst the Spanish. The successive threats and uprisings by the Chinese resulted in a total breakdown in relations between the two communities. In 1682 Don Francisco Diego de Aguilar

---

[97] Filipinas, 64, N.1, *Registros de champanes y pataches llegados a Manila*. 1 leg.
[98] Hagi Uragami Museum, *Trade Ceramics Found in the Philippines*, Exhibition Catalogue, Hagi, 2000, pp. 9-16.
[99] AGI Filipinas, 35, N.82, *Testimonio del numero de sangleyes que entran en Manila*. Fols.1226r-1227v
[100] AGI Filipinas, 35, N.82, *Testimonio del numero de sangleyes que entran en Manila* .Fols.1226r-1227v
[101] Antonio de Morga, *op. cit*, p. 187.
[102] AGI Filipinas, 27, N.148, *Petición del Cabildo secular de Manila sobre Parián de sangleyes.* Fols.887r-892v
[103] Antonio de Morga, *op. cit*, p. 187.
[104] AGI Filipinas, 29, N.57, *Carta de Francisco de las Misas sobre varios asuntos de Filipinas, comercio, salario*. Fols.382r-404v.

commented: 'Now the sangleyes, who are malign, dangerous and harmful to this republic of Manila, in the first place repeated riots... the case of many gambling that exist in Parián and in the extende province of Tondo, it is permitted and tolerated by the mayors who govern the said jurisdiction in return for the low interest which is given to the mayors, sometimes 100 or 200 pesos every month only for a gaming licence... The sangleyes are not necessary in this place for the supply and service of the labour they offer, since all... is so readily done by the natives, mestizos, Japanese and Ternates, who undertake their duties well and with loyalty.'[105] What the Spanish frequently accuse the sangleyes of is that they are people with no faith and who cause spiritual harm to those natives who are baptized.[106] A contrary opinion was given by Don Diego Calderon y Serrano in 1682, that even though they are not necessarily loyal, the quality of the work done by the *sangleyes* is incomparable. He goes on to make a strong case for their from the republic.[107]

A list exists of the trades of those *sangleyes* residing in and outside the city in 1695, and it helps us understand the social condition of *sangleyes* in the 17th century.[108]

1. Merchants with much business in trade, 10.
2. Shopkeepers selling textiles, white cloth, buttons, etc., 85.
3. Silversmiths in Parián and other towns around Manila, 18.
4. Tattooists with shops, 9.
5. Ironsmiths with forges and shops of all kinds selling iron goods, locks, padlocks, forks and tools for outdoor use, 27.
6. Bell-ringers, 2.
7. Christian officers, 7.
8. Tailors (with the labour being done by natives), 2
9. Rice-wine sellers, 5.
10. Embroiderers (with the labour being done by *mestizos* and natives), 1.
11. Painters and pavers (with the labour being done by mestizos and natives), 1.
12. Goldsmith (the labour being done by a *mestizo* because he is old and blind), 1.
13. Carpenters, 14.
14. Stonemason, (with a *sangley* as head worker), 1.
15. Silversmiths (all natives from Camarines and Pangasinan) (no number provided).
16. Boat constructers and house constructers (all natives) (no number provided).
17. Fisherman, 1.
18. Locksmiths with shops, 8.

The above table gives us a clearer image of the *sangleyes* in Manila. Most of them were probably originally labourers and by this time the actual work had been passed on to the natives and *mestizos*, while the *sangleyes* became shop owners and employers, probably climbing up the social ladder. It is true in a sense that Chinese labourers were no longer needed by Philippine society as *mestizos* and natives had taken over much of the work that used to be carried out by the Chinese in previous times.

The general image of the *sangleyes* in Manila is one of migrants engaged in trade with Guangdong or Fujian, but from the above table the numbers seem limited, although those *sangleyes* in Manila probably did the majority the more significant negotiations with the Chinese merchants. Further research needs to be undertaken, however, as it is possible to detect that these merchants were limited to a number of wealthy families with a strong family and business ties with southern China. These Chinese may be classified into three groups, depending on how they are related to the Philippine Islands:

1. Chinese who came to trade and returned as soon as their business was done.
2. Those that came to the Philippines and stayed for longer or shorter periods of time. Their main objective in this case not being trade.
3. Those who converted to Catholicism and stayed on the islands for the rest of their lives. Some were married to local women and obtained official citizenship.

Individuals in the first group of Chinese were frequently called 'captain' in many documents, and the above-mentioned study by Atsuko Hirayama refers to a Fujianese wealthy family tradition that supplied ships and capital to relatives and agents to facilitate trade. Not all 'captains' were ship owners, although some may have been substantial traders politically and financially connected to southern China, especially a group of merchants called 'anai' or 'anhai,' for which we have no Chinese character match at present.[109] These merchants with connection to China might have been the true threat to the Spaniards in terms of possible insurrection and for being potentially powerful leaders of a Chinese population twenty times larger than the Spanish. The countermeasure to this threat was strict segregation, whereby no Chinese were able to reside overnight in the *intramuros*. On the other hand, within Parián, the Spanish were more flexible, giving a certain level of autonomy, especially to Catholic Chinese citizens maintaining their own cultural traditions.

Regarding status and relationships between the Portuguese and Chinese in Manila, the Portuguese were focused on bringing products that would sell well in the

---

[105] AGI Filipinas, 28, N.131, *Expediente sobre expulsion de sangleyes*. Fols.960r-1130v.
[106] AGI Filipinas, 28, N.131, *Expediente sobre expulsion de sangleyes*. Fols.960r-1130v
[107] AGI Filipinas, 28, N.131, *Expediente sobre expulsion de sangleyes*. Fols.960r-1130v
[108] AGI Filipinas, 28, N.131, *Expediente sobre expulsion de sangleyes*. Fols.960r-1130v.

[109] Atsuko Hirayama, *op. cit*, p. 285.

American markets. Luxury goods were the major items, and most of the cargoes were probably trans-shipped directly to Acapulco. On the other hand the Chinese concentrated on goods for daily consumption in the Philippines and, thus, it may be inappropriate to assert that they were 'rivals'. Their aims and product ranges were different, as can be seen from some Zhangzhou ceramics imported by Chinese merchants from the late 16th century. They were apparently of inferior quality, even perhaps substitutes for Jingdezhen wares, either to be used locally or exported as cheaper items.

Ultimately, however, the Manila galleon trade, whose destined market was Spanish or *mestizo* consumers in New Spain, was implemented by Portuguese merchants more familiar with European markets and who also had access to extensive trade networks in Asia, and, in a way, who undertook their trade as agents of the Spanish.

Chapter II

Commerce and Merchants in the Manila Galleon Trade

1. Flows of goods from Manila to New Spain

By way of the Manila galleon trade there were flows of many Asian products and populations. What is important to discuss in this chapter is that not only textiles and silks were exported to New Spain but porcelain, Japanese lacquer-wares and items such as folding screens, which influenced American material culture. Slaves were also exported from Asia, further impacting on social and cultural levels. These cultural exchanges form part of the present cultures of Latin America and it needs to be discussed how Asian culture was accepted in society and how it was integrated over the course of time.

Who carried out this trade and caused this cultural exchange is of the same importance as considering the actual trading itself. In terms of the Manila galleon trade, Mexican merchants were the major investors in commerce and the number of tradeing merchants was limited to wealthy individuals and some religious orders. Among these the Portuguese *converso* merchants were prominent in that they used their own networks within New Spain as well as beyond America, via Europe, the Middle East and Asia, to carry out a large-scale trade. These people were of social and economic importance, especially in Mexico where the Inquisition by the Holy Office was carried out. Their existence and importance will also be discussed in this chapter.

After Miguel López de Legazpi settled in Cebu, one of his first intentions was to send a galleon with goods acquired in the Philippines. The future governor of the Philippines had to convince in any way he could that the conquest of the islands in the East was beneficial and would deliver real gains to Spain and New Spain. One of the earliest galleons, the *San Pablo*, was sent to New Spain from Cebu in 1565, with 200 seamen and a small quantity of cinnamon acquired in Mindanao.[110] In 1567 another galleon was despatched from Cebu to New Spain. In the succeeding year, 1568, a ship set out for New Spain although it is reported that 'it was lost in the islands of Guam which is one of the [islands] called Ladrones,[111] where a storm came and [the ship] was lost... [The] loss was great for us since the ship carried a large quantity of cinnamon and other products in order to give satisfaction to the Kingdom and the nobles of Your Majesty. The ship carried 150 quintales[112] of cinnamon, and as a personal cargo a further 250 quintales of cinnamon... [As] a sample, permission was given to load silk and porcelain and other curious products to content... the vassals of Your Majesty...'[113] Legazpi was also desperate for military support for these islands and which at last was offered from New Spain in 1573, after the founding of Manila as capital of the Philippines. The ship carried '150 soldiers by the order of Alonso Velazquez'.[114] However King Philip II must have doubted the profit Spain or New Spain might gain from the trade since the first cinnamon from the Philippines was so poor in quality.

Other products considered as profitable were 'pearls from the island of Ibabao and the island of Batayan and from Cagayan and Bohol and Mindanao between Cavite and Labaya of Baguindanao and the island of Joloc [Jolo] in [large] quantities. There are spices and drugs, especially in the island of Mindanao and Cavite and in Cagayan and Compoc and Taya together with Butuan. [Black] pepper, round, although... not so large in quantity, but there will be a certain quantity if they sow the seeds as much as they wish, such as in Cauchiu [Cochin] which is close to China. There are also gingers, tamarinds and other drugs without touching the coast of China, Siam, Patani. Neither in Java, nor Malucos, are there all these spices and drugs, and some precious stones exist according to the Portuguese who trade with Burney.'

The early phase of trade between Manila and New Spain was not as profitable as one might imagine, and in exchange for some products from the Philippines the major cargo sent from New Spain seems to have been soldiers.[115] In 1570 a galleon named *Espiritu Santo* sailed from Panai for New Spain on the order of Miguel Lopez de Legazpi. The cargo registered was black peppers, bought by the *Real Hacienda* (National Treasury), and other materials such as mast timber, bronze, iron, lead, wax, etc. On arrival, Rodrígo Espinosa, a galleon master pilot registered two porcelain jars, which were brought by two Portuguese named Domingo and Lucas. Other registered goods included a Chinese trunk, owned by one Hernando Riquel, with a piece of silk and some porcelain to submit in Mexico City. These goods were to be submitted to a factor called Gordian Casasano.

---

[110] Antonio M. Molina, *America en Filipinas*, Colecciones MAPFRE, Madrid, 1992, p. 96.
[111] The islands of Guam were called *Islas de Ladrones* (meaning thieves) by the Spanish, since the natives of these islands were so interested in iron and traded any food with pieces of iron.
[112] *Quintal* = one ton.
[113] EI. C-7/2/7, N.43, *Cartas y relaciones de las oficiales de Filipinas sobre la llegada y cerco de los Portugueses*.
[114] AGI Filipinas,35,N.3, *Carta de Juan Nuñez sobre su pobreza*. Fols.88r-96v
[115] Antonio Garcia-Abasolo, 'La llegada de los españoles a Extremo Oriente y la colonizacion de Filipinas', *Gran Historia Universa*, Vol. XXVII, *Descubrimiento y Conquista de America, Club Internacional del Libro*, Madrid, 1982, p. 60.

Another Chinese trunk belonged to Julian Mercado and contained green damask and six pieces of coarse cotton cloth from Luzon. Other notable goods were more than a dozen porcelain bowls, 20 pounds of linen and six porcelain vesselss to be submitted to a merchant who resided in the Mexico City, probably a black slave (who is only mentioned as 'black' and belonging to one of the Portuguese), 200 alfalfa, 300 large porcelain vessels, and 20 coarse cotton cloth pieces from Luzon. Another Portuguese, named Felipe, brought 700 items of porcelain – a good quantity.[116] From this document it can be observed that silk was not yet a major export to New Spain, whereas 300 porcelain vessels were submitted to a Mexican muleteer named Pedro, and 700 more were brought by a Portuguese called Felipe[117], are noteworthy. The Portuguese, who already had many decades of trade experience in Asia, could easily acquire goods in China and bring them to the Philippines via Macao. Although they may not have had any contracted merchants as recipients in Mexico, or other large cities, such as Puebla or Guadalajara, it is not difficult to imagine that they brought the goods themselves and sold them on in Mexico City at the markets and shops.

The intention of the Portuguese in the Manila galleon trade was to participate in a direct manner, competing with the Spanish. In 1590 a ship from Macao, called the Nuestra Señora de Asuncion, arrived at Acapulco with merchandise without license and all the goods were confiscated.[118] In 1598 the Spaniards noted that the Portuguese *Consejo* (council) ordered a ship be sent from Peru to Asia and then back to Portugal. It is recorded that, 'In the letter of 1590, which shows a desire to launch a strong artillery attack on the behalf of Gonzalo Fernando de Serrera, Diego Martinez Salazar and others. It may be possible to send a ship to China and return loaded with merchandise obliging it to bring 500 *quintales* of refined copper… This contract is prohibited from the time of the government of viceroy Don Martin Enriquez and it has been left aside, and Your Majesty was pleased to respond to the order since the ban was re-issued. What later occurred was that the ship made a voyage and arrived in Macao in 1591, and asked for the licence to enter the port [of Cavite].'[119] It is true that, from 1580, Portugal was under Spanish rule until its independence in 1640, although in case of commerce, especially relating to America and Asia, the Spanish government forcefully intended to prohibit Portuguese participation to prevent them from gaining benefits from trade. In 1592, conflict between the Spanish and the Portuguese authorities led to an agreement to cease direct trade between Macao and New Spain and that the Portuguese merchants should invest in the Pacific trade through Spanish or Portuguese agents in Manila.[120]

Detailed information on goods exported from Manila to Acapulco can be found in several written sources. A list of registered goods from 1581, certain cargoes belonging to Francisco Sande, is documented in detail and helps us to understand the products exported from Cavite.[121]

It contains the information that, 'In the last year of 1580, Francisco Palao and Don Bruno de Sanoe, on behalf of Doctor Francisco de Sande, who was the governor of these islands [the Philippines], in order to satisfy those who were jealous of the value of this merchandise, hired officers on the orders of Royal Court to investigate these products and sent judges, recipients at the port of Acapulco and other prosecutors to [inspect] the said products and merchandise...' The contents of these cargoes belonging to Francisco de Sande were:

- 22 boxes of white silk in bundles.
- 2 boxes of damasks.
- 1 box of woven silk with damasks inside.
- 20 boxes of raw silk.

Some of the above cargoes were again registered on entering Mexico City by a muleteer named Alonso Serrano, who claimed that he had brought 14 boxes and one bucket with clothes, which were brought for others and submitted to Sebastian Vazquez. These were:

- 8 boxes of ceramics decorated in gold.[122]
- 1 bucket of flat ceramics decorated with gold.
- 1 box of raw silk.

Another muleteer, Diego López, submitted to Bartolorme Sánchez de Molia the following goods:

- 7 packs of cloves.
- 14 packs of cotton thread.
- 1 pack of fine blue muslins.

Muleteer Antonio de Aragon declared:

- 6 sacks made from course cloth with black peppers.
- 8 packs of fine blue muslins.
- 1 pack of fine white muslins.

---

[116] AGI Contaduria 1196/1565-1576, *Caja de Filipinas, Cuentas de Real Hacienda*.
[117] In the primary source it is only mentioned as 'un Portugués' (a Portuguese).
[118] AGN General de Parte, Vol. 4, exp. 339. *Nao de Macan 1590*. Fol. 98.
[119] AGI Indiferente 745,N.168 Consulta del Consejo de Indias. 1 documento
[120] Goncalo Mesquitela, *História de Macau Volume II*, Instituto Cultural de Macau, 1997, p. 43.
[121] AGI Filipinas, 34, N.35, *Inventario de los bienes de Sande depositados por Diego López*. Fols.268r-275v.

Muleteer Alonso reported:

- 1 box of damask.[123]
- 1 box of gold decorated ceramics.

Muleteer Pul Vicente submitted to Diego López Montalban the following:

- 1 box of ceramics in gold and white.
- 1 vessel of ceramics in gold and white.
- 74 vessels of gold ceramics.

These lists show that the Spanish shipped large quantities of silk, porcelain and damasks as major export products to New Spain by 1580. Once they were registered on arrival at the port of Acapulco[123] and unloaded they were sorted depending on their ultimate destinations and recipient merchants, and were then carried by the muleteers hired by each merchant. Major destinations were Puebla, Guadalajara, Mexico and Veracruz. Since the journey from Acapulco to the inland regions was a succession of precipitous mountains and valleys, the muleteers had to carry the cargoes using mules or sometimes even porters.

The next detailed list of merchandise exported from Manila (1602) consists mainly of silk, taffetas, damasks, coarse cotton cloth and porcelain.[124] What follows is the inventory of Chinese goods sent by Alfredo Zuñiga to Antonio Rodríguez, a resident of La Puebla de Los Angeles:

Box 1

- With 100 pieces of black *gorbaranes*.[126]
- 20 pieces of taffeta.
- 8 pieces of black satin from Lanquin, rolled.[127]
- 6 *borcanses* and a half of silk from Lanquin.
- A piece of coarse cloth from Chincheo (Zhangzhou).[128]

Box 2

- 8 pieces of black Lanquin satin, rolled.
- 43 pieces of *gorbaranes* from Canton.
- 7 black *gorbaranes*.
- 3 large bundle of raw silk.
- another bundle of raw silk.
- 16 pieces of damasks of all colours.
- 2 bundles of raw silk.
- 29 pieces of coloured *gorbaranes*.
- 6 pieces of gauze for a canopy.
- 6 pieces of coloured gauze.
- 4 *catty*[129] and half of white silk.
- A bar of coarse cloth from Chincheo.
- A large bar and 2 bundles of raw silk.

Box 3

- Pieces of rolled black satin from Lanquin.
- 8 pieces of damask.
- 8 catty of white floral silk from Lanquin.[130]
- 15 pieces of white silk.
- 15 pieces of white satin from Lanquin.
- 43 pieces of black *gorbaranes* from Canton.
- 4 bars of raw silk from Lanquin in 24 bundles.
- 11 pieces of black *gorbaranes* from Canton.
- 6 catties and half of white floral silk from Lanquin.
- 9 pieces of black *gorbaranes* from Canton.

Box 4

- A coarse cloth from Chincheo.
- 19 pieces of rolled black satin from Lanquin.
- 18 damasks of all colours.
- 16 pieces of white satin from Lanquin.
- 8 catties of floral white fine Lanquin silk.
- 3 bars and a half of raw Lanquin silk.
- 23 coloured *gorbaranes*.
- 21 pieces of black *gorbaranes* from Canton.
- 12 catties of white floral silk from Lanquin.

Box 5

- 67 pieces of coloured *gorbaranes*.
- 30 pieces of white Lanquin satin.
- 4 large bars of raw Lanquin silk.
- 13 catty and a half of floral Lanquin silk, fine and white.
- 23 black *gorbaranes* from Canton.
- 1 large bar of raw Lanquin silk.
- 10 pieces of black *gorbaranes* from Canton.
- 4000 pieces of ceramics.

It is clear that at this time most of the cargoes were of Chinese silk, substantiating previous studies of the Manila galleon trade. However, we must not simply conclude that this trade was for the mere exchange of silk and silver.

---

[122] These are actually blue-and-white Jingdezhen porcelain vessels with gold decorations.
[123] Acapulco was known to be a miserable and unhealthy village, of only about 20 Spanish houses, and only used as a port when ships came in from the Philippines or Peru. Several accounts can be found from various travellers, one of which is by Fransico Carletti, in his famous journal *Mi Viaje Alrededor del Mundo*, Editorial Noray, Barcelona, 2006. p.72.
[124] AGN Indiferente Virreinal 4976-0006, *Mercadurias*.
[125] The author has omitted the prices of these goods as they changed over a short period of time depending on supply and demand.
[126] 'Lanquin' probably indicates the city of Nanquin, from where all the emperors of China have ordered their silk gowns, historically of the best quality.
[127] Zhangzhou is located in present-day Fujian province.

Chinese silk has always been an important and profitable traded commodity in Asia. Japan and Vietnam also produced silk, although the quality never reached that of southern China. It is not surprising, therefore, that the major export product from China to the America was Chinese silk. Adding to this, in Spain, New Spain and America, silk production was also very important; Andalucian silk in particular was of better quality and was traded for higher prices than cheaper Chinese silks, and other textiles became affordable to all social classes and thus gained much popularity. Nevertheless, it is also important to note that Chinese porcelain from Jingdezhen, and certain other articles, was also traded and distributed throughout America. In all the above-mentioned lists of goods received in Acapulco there were good quantities of ceramics – for instance among the imports received by the merchant Antonio Rodriguez in La Puebla there were 4000 items of Chinese ceramics, a quite considerable amount, bearing in mind the fragility and distrance involved in terms of one single distributor. Furthermore, it is of some interest that all these pieces were destined for New Spain, a market that previously had no history whatever for the consumption of porcelain. In only around three decades, or less, Chinese porcelain was accepted and distributed in large quantities in America. Other decorative objects, such as folding screens and lacquer wares with mother-of-pearl inlays, were also exported from Manila. Folding screens were produced in Japan during the early Edo period and are often referred to as *biobu* in Japanese, and these were exported to New Spain. Some were highly-prized, especially those with scenes depicting Kyoto (the so-called *Rakuchu Rakugaizu* screens (with scenes of the interior and exterior of the city). These were often copied by Mexican artists, who called them *biombo* or *biobo*, and they became very familiar in Mexican society. Japanese folding screens were further exported to Spain and gained popularity in the highest circles, including the court. Lacquer wares were also exported, being copied in Mexico as *enconchado* (shell inlay). Since these goods were exported as luxury products, and were bought in smaller quantities, they do not appear in the lists of registered goods as frequently as silk. The memorandum of imports submitted in Acapulco by a merchant named Ascanio Guiajoni in 1628 includes a vague account of Japanese goods. The majority of the merchandise was taken up with Chinese silk, damasks, fine cotton and taffetas, although it also included '41 jars with 20 pico of black peppers', 'bedcover with embroideries from India', and 'a box full of different things from Japan'.[128] In a register of 1640, itemising goods from Acapulco entering Mexico City, there is mention of four *escritorios*[129] and a single *biobo* from Japan.[130]

Slaves were also traded to New Spain and these occasionally appear in documents.[131] In the petition of Captain Juan Lopez de Ando (1633), master of the galleon *San Juan Baptista*, we read: 'I, Juan Lopez de Ando, ask [the king's permission] to sell... to Simón López, a merchant who resides in this port... [a] Chinese slave named Domingo, who should be 20 years old, more or less'. The document further mentions that this Chinese slave is a captive and thus free to be sold, with no previous financial impediments, for 250 gold pesos. Simón López is a well-known merchant from Zacatecas who made his fortune from the Manila galleon trade. He was also known as a rich Portuguese *converso* and was later brought before the Inquisition in Mexico.

The slave trade was largely carried out by the Portuguese within Asia and there are accounts that even Japanese, Korean and Chinese salves were bought by Portuguese merchants, via Jesuit intermediaries, and mostly sold within Portuguese colonies in Asia.[132] Among the cargo of Luis Lobo Costilla entering Mexico City in 1640 there was 'one Chinese slave' registered with the other merchandise. This trade in Asian slaves to America is a topic little studied, and as not many appear in the written sources it is difficult to know approximate numbers. It is not clear whether such slaves were actually widely 'traded' or were the possessions of Spanish officers and merchants. It is even difficult to define the term 'slave trade' within the context of the Manila galleon trade since there are no records known of an organized trade through this route, compared to the huge numbers of black slaves brought over from Africa by the Portuguese. Slavery in some form already existed long before the arrival of European powers in Asia. The definition of slavery in the Philippines was already being argued among Spanish authorities by 1573.[133] This indicates that there were slaves in the Philippines, some born into slavery, some as prisoners of war, and some from debt. Thus many Spaniards were able to purchase these slaves and put them to work as servants in their homes.[134] It seems also true that the Spanish authorities considered these slaves to be a potential source of manpower for American mines. This might have led to the mass export of slaves to New Spain; the governor of Manila went so far as to confirm that these slaves were note required in the Philippines to work in agriculture and supply food. It seems however that the slave trade was never operated by slave merchants as such, nor that there was an organized slave trade in the same way that the Portuguese

---

[128] AGN Indiferente Virreinal, caja-exp.: 4259-012.
[129] *Escritorio* is a Spanish term for 'desk'.
[130] AGN Indiferente Virreinal, caja-exp.: 6441.
[131] AGN GD64, *Jesuitas*, Año: 1581-1645. Vol. IV, Fol,50.
[132] Michio Kitahara, *Portugal no Shokuminchi Keisei to Nihonjin Dorei*, Kodensha, Tokyo, 2013, p. 14.
[133] AGI Filipinas, 6, R.2, N.16, *Carta de Guido de Lavezaris sobre los esclavos de Filipinas*. 1 documento.
[134] Slavery in the Philippines was prohibited by the Spanish authorities in 1574, although the Spanish continued to buy and have them serve in their houses, or have them accompany them on voyages and travels.

undertook it between America and Africa, although there were slaves sold in Acapulco by their owners.[135]

The population of Chinese immigrants grew quickly in America, some voluntarily and some as slaves. In 1635, the Mexico City municipal council received a petition to reduce the number of Chinese barbers to twelve, and that they should be located in the suburbs to avoid conflict with Spanish barbers. It is noted that in 1635 there were more than 20,000 Chinese in Mexico City and many of those working as barbers arrived in New Spain as slaves.[136] In one of these documents it is mentioned that in 1637 186 slaves arrived in Acapulco and all the owners were ordered to pay 50 pesos to each of their slaves.[137] This sum is probably a wage rather than being a market price. In the port of Acapulco, along the coast, there was a group of Asians identifying themselves as 'Chino' or 'Indio-chino', among them there were 21 Filipinos (Manila, Cebú, Pampanga, Cagayan, Zambales), 5 Indians, 2 Ternates, and 2 Papuas. 52 of the Chinese immigrants identify themselves as 'slaves', the majority being labourers who worked in haciendas or were servants. There was a difference in value between African and Asian slaves, with African slaves being more prized than Asians. The valuation difference can be compared in the property register of Nuestra Señora de la Concepción, where it states that Africans aged between 20 and 30 are valued at approximately 400 pesos, while 'Chinos' are valued at 300 pesos in the latter half of the 16th century and until 1700 in Coyuca.[138] A document dated to 1640 records '3 Chinese slaves' registered as being submitted to Alonso Beato de Rojas along with other Chinese merchandise.[139] It is true that some Asian slaves were labourers, exported to New Spain via Manila, although it is unclear whether we should define this as trade as such. This might have been closer to a form of migration process, with transported from one place to another with their Spanish owners. Although there appears to have been no organized slave trade identical to that carried out by the Portuguese, the cultural influence of Chinese 'migrants' to Mexican society should not be underestimated.

The commodities we have been discussing above are focused on imports from Manila to New Spain in return mainly for silver, although the return cargoes to Asia included other necessities, such as arms and munitions, refined saltpetre, lead, wine, wheat-flour for bread, including the host used in the Catholic mass, olive oil, vinegar, threads and rope, sails from The Netherlands, white paper from Geneva, books, mostly of European provenance, etc. Comestibles included lard and preserved meats, cheeses, fruit and nuts.[140] The foodstuffs it seems were mostly for consumption on the long voyages and for Spanish residents in the archipelago. Other notable items sent to Manila from Acapulco were, as mentioned above, soldiers, money, munitions and goods to sustain the lives of Spanish officers and members of the religious orders. Soldiers and arms were in constant demand in the Philippines owing to the small Spanish population and the constant tensions between the large Chinese community in Manila, and the presence of Dutch, English and, later, French vessels in Asian waters. Once the merchandise from Manila had been received the payment would then be sent to the Manila merchants. Other trade mechanisms existed, such as that operated by Diego Rodrígo de Torres. This merchant had estatblished a company with a carrier, Lorenzo de Figueroa,[141] who went to Manila as agent and bought goods there and sent them on to Mexico. The money gained by selling these goods in Mexico would be sent back to de Figueroa as working capital for the next consignment.[142] Most of the products exported to New Spain came from China, especially from the south of the country. Silks were produced in Hangzhou and Nanjing; they were shipped from Fujian or sold in the market of Guangdong. Porcelain was produced in Jingdezhen (Jiangxi Province) and transported to the port of Guangdong or to the market there. Japanese goods were probably produced near Kyoto, although the exported folding screens and lacquer wares were mass produced and of inferior quality, and their exact places of production are unknown.

2. Market structure and Mexican merchants

From the early phase of the Manila galleon trade many merchants participated so as to profit from it, including Spanish, Portuguese, Mexican and Peruvian traders, among others. The position of the Mexican merchants strengthened in the 1570s after the start of the Manila galleon trade. The Spanish authorities were diligent in protecting their wider interests and were quick to stipulate how trade should be carried out. Spanish influence grew quickly as a result of significant flows of coin to China (Chinese textiles were only bartered for gold, silver, or coin). During this early phase, particularly in the early 1580s, commerce between New Spain, Peru and Manila rapidly developed without restriction. However the unexpectedly large flows of coin had to be controlled and Philip II prohibited trade between

---

[135] AGI Filipinas, 340, L.3, *Orden sobre esclavos de pasajeros en las naos de Filipinas*.
[136] H. Homer and Robert S. Smith, 'Chinese in Mexico City in 1635', *The Far Eastern Quarterly*, Vol.1, No.4 (Aug 1942), pp. 387-389.
[137] AGN GD100, *Reales Cedulas*, Vol. 11, exp. 451, Fol. 317.
[138] Déborah Oropeza Keresey, *Los 'indios chinos' en la Nueva España; la inmigración de la nao de China, 1565-1700*, El Colegio de México, Ph.D. Dissertation, 2009, p. 98.
[139] AGN Indiferente Virreinal, caja 6441, exp.: 092 (1640).

[140] Oswald Sales Colín, 'Las Cargazones del Galeón de la Carrera de Poniente', *Revista de Histórica Económica*, Año XVIII, Otoño-invierno, pp. 645-646.
[141] The term 'cargador', which literally means 'carrier', may have been used to indicate agents.
[142] Guillermina del Valle Pavón, *Los Mercaderes de México y la transgresión de los limites al comercio pacífico en Nueva España, 1550-1620*, p. 229.

Manila and Peru. In 1583 Spain strongly protested that excessive quantities of Chinese silk were being imported to New Spain and distributed throughout America, and consequently Spanish silk (especially Andalucian silk) had to compete with cheap Chinese textiles. Spanish trade was thus strictly controlled by the authorities, based on protectionism, although of course it was always difficult to control every trade movement far from Spain. Even the existence of the *Casa de Contratación* (founded in 1503), which was the first organization set up to control all Spanish commercial interests, including trans-Atlantic trade and communication with all Spanish ports, both internal and external, receiving reports of their dealings and the situation of the colonial markets,[143] did not succeed in the control of trade in the far-flung colonies.

Although the Manila galleon trade is frequently seen as the driver of prosperity for the Philippines, especially Manila (as a cosmopolitan trading port), the real benefit from commerce went to the Mexican merchants. In 1593 Philip II issued an order that only residents in the Philippines were able to undertake trade and prohibited them from acting as wholesalers to the Mexicans. This move was designed aid the poorer Spanish traders in general in the Philippines.[144] The *repartimiento* system, a means of distribution of galleon lading space for the transporting of merchandise, was intended to enable citizens of the islands to participate and profit from the traffic. However, as a consequence, commercial power became increasingly concentrated in the hands of a consolidated merchant class. The authorities went on to set a limit of 250,000 pesos on the value of merchandise that could be sent to Acapulco.[145] Furthermore, ships were to be despatched twice a year, carrying 300 tons each. However, it is generally accepted that this regulation was never adhered to, and facilitated volumes of unregistered shipments that were simply unloaded before reaching Acapulco. In 1595 and 1596, merchandise worth 580,000 pesos was sent to America by 437 individuals, 30 of whom were members of the *consulado*, in Mexico only 31 individuals paid for the licence.[146] The *consulado* functioned from the late 16th to the late 18th centuries and was a group of merchants controlling the commercial activities of New Spain. Members of the *consulado* were able to send their agents to Manila to undertake the necessary arrangements for the provision of merchandise needed in Mexico.[147] In fact, regardless of all the regulations imposed by the *repartimiento*, and the incentives for those merchants residing in Manila, the survival of the Philippines within the Spanish empire depended on Mexican investments and the direct interest of colonial officials and merchants in New Spain. As a consequence, the Manila galleon trade continued in spite of all of the opposition of the metropolitan merchants.[148] This state of affairs was probably the result of the fact that trade from Manila was administrated directly by New Spain and was able to escape Spanish control.

Most of the merchandise was destined for Mexico City, although, as previously mentioned, La Puebla, Guadalajara, Oaxaca, and Veracruz also had markets and were pleased to receive Asian goods. In Mexico a licence was needed to sell Chinese, Spanish and local goods, as mentioned in a written source: 'In Mexico City, 27 July, 1651... a licence will be issued and approved to sell goods in the road... [Merchandise] from Castile, China, and from this land'.[149] The largest market in Mexico City was located in the centre of the city, now called Plaza Mayor (where stand the main Cathedral, the National Palace, and the City Hall) and which was, and still is, the political and religious centre of the metropolis. There were numerous small shops/wooden stalls (*tianguis*) where the majority of smaller items and foodstuffs were sold. According to an account by Thomas Gauge, who went to Mexico City in 1625, 'the most important plaza in this city is the market which, without having larger scale than it had in the time of Moctezuma, is huge and very beautiful. One of the sides runs in the form of a portico, or arcades, under which there is shelter [from rain]. The plaza is occupied by shops of merchants selling silk and clothes of the most varied assortment... [There] are stands of women selling spices, fruit and herbs... [Higher value] merchandise includes salt, coarse cotton cloth of diverse colours and size, for cushions, beds, clothese... and [there are] trinkets for the home. There are also pearls, precious stones, shells of various types, sponges, and many more.'[150] Gauge also refers to specific streets, i.e. 'the street of San Augustín is also very rich and pleasant, where silk merchants live.' These merchants Gauge refers to were wholesalers of Chinese silks. Once these merchants received goods from Manila, they would sell them on to the retailers who would put them on offer at the Plaza Mayor. In the 18th century the Plaza Mayor began being referred to as 'Parián', which derived from the Chinese residential area in Manila.

According to the regulations, eight residents of the Philippines were able to send merchandise officially to New Spain, although many of these residents were

---

[143] Clarence H. Haring, *Comercio y Navegacion entre España y las Indias*, Fondo de Cultura Económica, México, 1939, pp. 27-57.
[144] AGI Filipinas 77, N.6, *Carta del Cabildo eclesiástico de Manila sobre comercio de Filipinas con Nueva España*. 2 documentos.
[145] William Lytle Shurz, *The Manila Galleon*, Dutton Everyman Series, New York, 1959, pp. 155-157.
[146] Guillermina del Valle Pavón, *op. cit*, pp. 227-228.
[147] Guillermina del Valle Pavón, *op. cit*, pp. 218-228.

[148] Katherine Bjork, 'The link that kept the Philippines Spanish: Mexican Merchant Interest and the Manila Trade, 1571-1815', *Journal of World History*, Hawaii University Press, Vol.9, No.1, Spring 1998, pp. 25-30.
[149] AGN Archivo Historico de Hacienda, leg. 268, exp. 196.
[150] Thomas Gage, *Nuevo Reconocimiento de las Indias Occidentales*, Fondo de Cultura Economica, México, 1982, pp. 159-186.

agents hired by Mexican merchants, or, in many cases, were passengers, sailors and captains who were allowed to bring in merchandise as their own private property, and thus avoided paying the *almojarifazgo* (customs tax). The tax imposed was 3% when sailing out from Cavite and another 3% when entering the port of Acapulco, and finally 10% at the entrance to Mexico City. Regardless of regulations, Pacific trade was mostly invested in by the small group of wealthy Mexican merchants who sent capital and agents to Manila. The study by Luisa Hoberman provides an interesting note of the differences between large-, medium-, and small-scale commercial activities from the late 16th until the middle of the 17th century.[151] Figure 2 shows the tax paid and the percentages relating to the various merchants. It is clear that the larger merchants paid most in taxes, especially from 1630 to the 1690s. Figure 3 indicates the decrease in the number of merchants, and even though large merchants were fewer in number they were actually paying most of the tax, meaning that this minority gradually became the key to the Mexican economy.

In general, merchants passed their wealth on only to their eldest sons, and the other children were sent to religious orders. In other cases, when a merchant died all assets would go to the widow and if she remarried the wealth would transfer to the new husband, who was usually also a merchant. Thus the merchant wealth in New Spain was, in most cases, left undivided between their children, and rich merchant families stayed wealthy for centuries.[152]

Trans-Pacific trade was especially attractive to Mexican merchants, as a chance to acquire fortunes, whereas for Manila merchants it was seen more as an economic lifeline. Even with the royal decrees issued to protect and improve the lives of those merchants residing in Manila and prohibiting Mexican merchants from intervening directly in the trade, the Mexican merchants continued to send their agents to the Philippines and inflicted losses on the Manila merchants. In 1632 a petition was raised to protect the monopoly the Filipino merchants had regarding island trade, which it seems was still being damaged by those Mexican merchants who had recently delivered large quantities of money to the islanders to purchase goods for them that would return in the personal baggage of sailors and officers, resulting in surpluses and tax evasion. The document also refers to the absolute prohibition of Mexican agents living and acting as merchants in the Philippines,[153] although this order was never followed strictly.

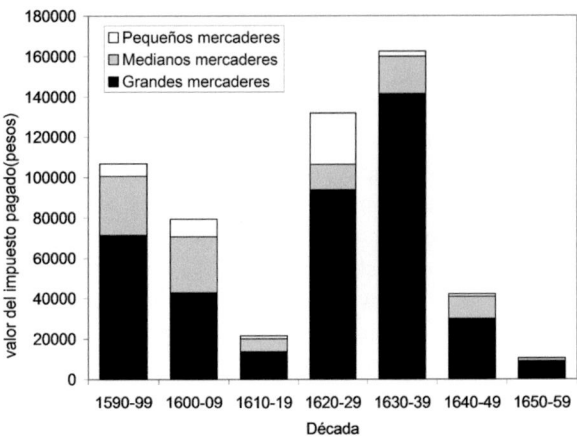

FIGURE 2: TAXES PAID BY MEXICAN MERCHANTS.

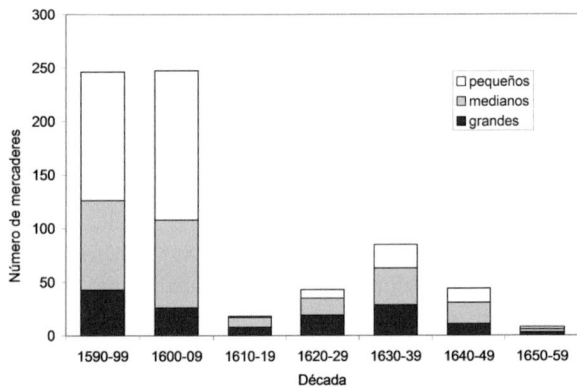

FIGURE 3: NUMBER OF MERCHANTS PARTICIPATING IN TRANS-PACIFIC TRADE.

The participation of Asian merchants is another issue to discuss. An example is a source referring to an Asian merchant named Tomás Pangasinan (described as *chino-indio*), who brought Chinese textiles all the way to Acapulco himself. The document is dated 1591 and shows that already at this stage, Chinese/Asian merchants were participating directly in this trade. Their participation may have been modest, although it is worthy of further research since it relates to Asian migration to America, as well as to the activities of Asian merchants in the Manila galleon trade.

3. The participation of religious orders and the issue of contraband

The Manila galleon trade was based on private and free trade, relying on merchants. Religious orders, and those associated with them, were not officially permitted to participate in trade. However on several occasions Jesuits and Augustinians were accused of commercial activities, possibly using their own networks based in Manila, Macao and Japan. Koichiro Takase has

---

[151] Luisa Schell Hoberman, *Mexico's Merchant Elite, 1590-1660? Silver, State, and Society*, Duke University Press, Durham and London, 1991, pp. 40-41.
[152] Etsuko Miyata, *El Comercio de Porcelanas Orientales en el Galeón de Manila durante los siglos XVI-XVII*, Trabajo de Investigación Tutelado, Universidad de Santiago de Compostela, 2007, pp. 79-80.
[153] AGI Filipinas, 27, N.156, *Varios documentos sobre proteger a comerciantes de Filipinas*. Fols.908r-922v

[154] Based on Hoberman's data and schematized by the current author.

conducted an in-depth study into Jesuit participation in the silk/silver trade in Japan. He mentions that the order in Japan participated in the raw silk trade, acting openly as merchants.[155] The priests not only dealt in silk, but also in gold, mercury, lead and other textiles. Their commercial activities were endorsed by their central authority so as to sustain their religious drive into Japan and Macao. These activities ran contrary to the ban on religious orders participating in trade during Spanish rule in the Philippines. Successive petitions to prohibit such trade exist, themselves indicating how frequently the Jesuits and others disregarded them. An early account of interventions in trade by sections of the established Church can be seen in a letter from Diego Aduarte (1619).[156] Aduarte complains that one archbishop and three bishops arranged to bring Chinese merchandise into New Spain, and the profits they made they re-invested in next galleon convoy. An inventory of 1640 lists a merchant named Antonio Pinel, a resident of Chilapa, who submitted two boxes from Manila into Mexico City, aided by a Jesuit priest named Alonso de Rojas. The latter in turn sent these boxes to another Jesuit, Tombro Gomez.[157] The same document mentions a cargo beloning to Luis Lobato, declaring 'one box sealed by wax…three chests from China…[and] one large wicker basket of ceramics'. All of which were declared under oath, confirming that they belonged to one 'fray Pedro de Quesada of San Agustín', and also a slave was brought for his service by order of San Agustin, being registered as 'one small Chinese [man]'.

Even if infrequent, the fact that religious orders were in some ways connected to trade can be attested to both historically and archaeologically. In 1673 Juan López wrote to the king that 'in 1671, merchandise arrived particularly for [clerics]. Your poorer clergies in these islands might have an excuse to exercise some small business in order to sustain their lives, which is a natural right, although there is no apparent excuse for [them] to extinguish a complete province by means of their [trading activities].'[158] In 1699 a council of the Cathedral of Manila wrote to the king soliciting him to prohibit trade by certain clerics who were receiving and distributing cargoes from the galleons coming from New Spain.[159]

None of these primary sources indicate in detail how these ecclesiastics intervened in trade. However, considering that Jesuit Asian interests (in Macao and, to a limited extent, in Japan), were maintained and carried out far away from its base in Portugal, it was inevitable that the order would become heavily dependent on its own trading activities. The same may have been the case with the Spanish orders in the Philippines, where there were no effective investments, i.e. plantation projects, or sufficient support from New Spain. Furthermore, easy access to Macao, in terms of human and religious networks, made it easier for clerics to participate in the import and export of Chinese silk, porcelains, and other goods from Asia in general. This theory can be substantiated by the finds of large quantities of Chinese ceramics excavated from the convents and monasteries of Mexico City. The convent of Santa Teresa, close to the Plaza Mayor, revealed abundant examples of blue-and-white porcelain from the late 16th and early 17th centuries. From the Archbishops' residence in Mexico City many Chinese blue-and-white wares have been found, including a bowl decorated with a two-headed eagle (Photo. 6). These may well have been specially ordered pieces on behalf of the Augustinian order, since its crest is represented by a two-headed eagle. The participation by religious orders in the Manila galleon trade may also be explained in terms of the previously mentioned structure of Mexico's elite merchant families, whereby only the eldest son could inherit and the other children would be consigned to convents and monasteries. These individuals may also have been connected to, or even collaborated in, the practice of trade in the Far East. However, in any case, how they built up these networks and how they functioned both require further research.

The issue of contraband is another important issue to address. Smuggling as related to the Manila galleon trade has been less well studied in its history compared to trans-Atlantic or Caribbean trade, where contraband and piracy were mundane events. Although primary sources on contraband are few in the 16th and 17th centuries, a few do exist.

PHOTOGRAPH 6: JINGDEZHEN POLYCHROME BOWL WITH DOUBLE-HEADED EAGLE FROM THE EXCAVATIONS AT LA CALLE LICENCIADO DE VERDAD ©INAH.

---

[155] Koishiro Takase, 'Kirishitan kyokai no Bouekikatsudo', *The Socio-Economic History Society*, Tokyo, 1977, p. 55.
[156] AGI Filipinas, 85, N.34, *Carta de dominico Diego Aduarte sobre comercio de Filipinas.* 1 documento.
[157] AGN Indiferente Virreinal, caja 6441, exp.: 092 (1640). It is unclear whether these items were personal belongings or presents.
[158] AGI Filipinas, 74, N.137, *Carta de Juan López sobre bula de comercio de eclesiásticos.* Fols1010r-1011v.
[159] AGI Filipinas, 28, N.160, *Petición del Cabildo secular de Manila sobre carga de naos.*Fols.1469r-1470v.

Doña Margarita de Alarcon	3 half-cakes of crude wax.
Don Diego de Chivelli y Chamis	2 boxes.
Don Diego de Castaneda	1/2 a box.
Don Pedro de Medrano	52 boxes, 9 half-cakes of crude wax, 2 half boxes.
Don Domingo Lizarral de Cael	14 boxes, 15 half-cakes of crude wax, 50 of black peppers (?).
Doña Josephia de la Alina	100 boxes, 11 half-cakes of crude wax.
Doña Josephia de la Alina (to her spouse)	90 boxes, 3 half cases.
Don Luiz de Moralos Juan Bautista Carocaes Diego Molero Martis de Jauriqin Domingo Gonzales de Lisboa Luis de Rivas	45 half-cakes of crude wax, 2 quarter boxes and 3 half-cakes of crude wax, 3 boxes, 5 boxes and another 8 boxes.
Augustio Hernández	5 half boxes.
General Antonis Nieto	23 boxes, 6 half-cakes of crude wax, 47 boxes.
Don Joseph López de Viscarra	31 boxes.

FIGURE 3: LIST OF CONFISCATED CONTRABAND.

In 1697, for example, unregistered merchandise was found in the port of Cavite.[160] A record refers to a number of boxes to be loaded at the port and shipped to Acapulco. Being unregistered they were duly confiscated by the authorities there. These comprised:

The dominant product in this list of contraband is wax destined for New Spain. The other items, vaguely registered as 'boxes' may be silk, although the exact contents cannot be confirmed. A shipment of 90 and 100 boxes for Doña Josephina de la Alina represent large quantities of personal belongings, even if 90 boxes were to be sent to her husband. It is more feasiblel to consider that these goods were for resale.

In 1646, one Francisco Carrión, a merchant and resident of Mexico City, was charged with depositing in a warehouse some textiles which were traded clandestinely. Other suspicious instances, such as an investigation into some cargoes belonging to Juan de Vargas Hurtado, a governor of the Philippines, and which may have escaped accurate registration, appeared in a further document dated 1685.

Trade in contraband might have been more frequent on trans-Pacific routes, since, after despatch from Manila, ships would cross the Pacific until reaching present-day California and then would steer south, cruising along the coast until Acapulco. All the coastal area of Baja California could have offered sites for the unloading of goods, although no records of smuggling in this region have surfaced to date.

---
[160] AGN Caja Real Filipinas 1251, *Los bienes sin registrado en Cavite 1697*. Fols.123-150

5. The merchant diaspora and networks

For the Manila galleon trade specifically large Mexican wholesalers were the key to successful commercial ventures and profited most. These 'Mexican merchants' included individuals of Spanish, Portuguese, Italian and other European backgrounds. Traditionally Italian merchants were active in Atlantic trade, using their Mediterranean networks to further their interests. The Portuguese were prominent in the slave trade and also participated in Atlantic trade towards north, towards Amsterdam and Hamburg.

Any discussion on 16th- and 17th-century Mexican society and commerce must inevitably include references to these Portuguese merchants, who were *conversos*, not because all of them were wealthy and had great influence on society, but because of their shared religion, which strongly linked them personally and economically, and also their extant diaspora that crossed the globe and differentiated them from other Portuguese, Spanish or European merchants. Of course, the Portuguese were more independent from controlling authorities in all regions of Asia, America and Africa compared to the Spanish, and once reaching places of Portuguese interest or possession many of them dispersed to wherever they could make any sort of a living, and thus many became itinerants, wandering from one place to another. This process may well have underpinned commercial networks and established a parrallel community to the Portuguese *conversos*. Many of these Portuguese itinerants were hired by local interests in Asia as mercenaries, or for other occupations, and may not have had substantial contacts with one another, whereas *conversos* intentionally constructed their own networks, using discreet religious understanding as a medium.

The 16th and first half of the 17th century witnessed the added complexity of an influx of Portuguese *converso* traders into Spanish colonies, particularly New Spain and Peru, and the subsequent persecutions of the Inquisition (especially in the 1640s). In this period the trading networks of these *conversos*, as well as their religious orthodoxy, were put under investigation, and surviving records spotlight how the merchants of the time operated their joint trade. In terms of riches, as pointed out by Jonathan Israel, not all of them were the wealthiest merchants of the time, and it is more fair to say that most were of humble origin and came to New Spain to seek ways of making a living or seeking opportunities in a new land where they could escape the Inquisition underway in the Iberian Peninsula. Their major activities were as merchants engaged in the slave trade (black slaves from Africa to New Spain), the import of cacao from Venezuela, and the textile trade from China via Manila. Their importance may well be emphasized within the context of Atlantic trade, although it is clear that among the known merchants, such as Simón Váez Sevilla, Matías Rodríguez Oliveira, and Francisco Texoso, some were actively participating in the galleon trade, buying mostly Chinese silk. Most of these merchants residing in Mexico City, Veracruz and Guadalajara were financially connected. For example, Simón López, who was a merchant mainly engaged in the slave trade, was a business partner of Simón Váez Sevilla and co-owned a store. Simón Váez de Sevilla was a wealthy merchant in New Spain and was a brother-in-law of Tomás Nuñez, also connected financially, who lent money to other *conversos*. When the rich merchant Juan Duarte Espinosa was charged of practising Judaism and had his properties confiscated, both of them were named as money-lenders in the document. Tomás was also in Simon's house when he was arrested by the Holy Office (*Santo Oficio*), where Simón had a synagogue. His father-in-law, Antonio Rodríguez Arias, also a *converso*, left from Seville to La Puebla in 1600 and had already became a large wholesaler of Chinese merchandise by 1603. Simón Váez Sevilla was born in Castelo Branco (Portugal) and was said to have commercial contacts with merchants in Manila and Mexico, and from there all the way to Pisa (Italy) where some of his family lived. Simón Fernández de Torres and Simón Suárez de Espinosa were charged by the Holy Office in 1645, both being in some way involved with the importing of lead from China. There were many other *conversos* engaged in both the Atlantic and Pacific trade.

Additionally, a great many individuals of Portuguese origin were charged by the Inquisition from 1620-1650 with practising Judaism in secret. Referrences to '*portugueses*' in New Spain in fact meant '*conversos*', as those who were charged by the Inquisition in Spain either converted completely or fled to Portugal, keeping their faith within a closed society of *conversos* and passed their wealth on to their children, who in turn migrated to the Americas. Their particular success can be partly explained by their networks that protected their secrecy and a worldwide diaspora in order to survive the constant threat of the Inquisition.

In the register list of cargoes entering Mexico City between 1635 and 1641, five *conversos* are listed among 50 merchants importing merchandise from Manila. Although they are not so numerous as the merchants and investors in the Manila galleon trade, some names of larger wholesalers, such as Simón Váez Sevilla and Tomás Treviño de Sobremonte, can be seen on this list.

Further study is needed to clarify the link between America and Asian *conversos*. Their activities in America in terms of links to the Asian market are very difficult to trace since these Portuguese *conversos* were acting within the territories of both the Spanish and Portuguese crowns. Even when the two kingdoms were united from 1580 to 1640, merchant activities in Macao and other Asian provinces were rarely documented in the Spanish archives. However there were many men who left Portugal and sailed to the Portuguese overseas possessions in Asia in an attempt to escape religious persecution, and there must have been *conversos* carrying out trade in Asia. One we know of was the prominent figure Bartolomé Vaz Landeiro. Landeiro was a trader in Asia serving the king as an authorized merchant as well as participating in the Japan–Macao trade. He was born in Santa Iria, a small village on the outskirts of Lisbon and came from a Jewish family. He undertook official trade between Macao and the Philippines in 1583 and was active between Macao and Manila. His trading activities are documented from Japan, Macao via Manila, Siam, Cambodia, Timor, India (and probably many other trading posts which were under Portuguese control).

Chapter III

Exported Chinese Porcelain in New Spain

1. The export route from southern China to New Spain

As was mentioned in the previous chapters, Chinese porcelain produced in Jingdezhen was one of the important exports for the Manila galleon trade. Wonderful thin-pasted, blue-and-white wares containing a high percentage of kaolin did not exist anywhere else at that time and must have attracted the markets in the Americas just as it did everywhere else in the world. However it must be remembered that Chinese ceramics were only one of the many Asian products exported to New Spain. Silks (raw or in textiles), needless to say, were the most important products handled by the Manila galleon trade, although they do not survive the passage of time. For this reason ceramics are perhaps the most useful means for helping us reconstruct early trade routes, spatial distribution, and the nature of local markets in the past.

Manila simultaneously became the major trading platform and intersection point of not only Chinese silk but also porcelain coming from southern China. Some quantities were re-exported to New Spain and others were consumed in Manila or elsewhere in the Philippines. It is also important to confirm that Chinese ceramics were not 'new' to Philippine markets when the Spaniards came and settled in the 16th century. In the long history of the pre-colonial Philippines there was always a need and certain levels of consumption of Chinese ceramics – at lease as early as the 9th century Tang dynasty, as researched by several archaeologists and art historians such as H. Otley Beyer, Olov R.T. Janse, Robert B. Fox and Yoji Aoyagi. From the 9th to the beginning of the 15th centuries, burial sites in Luzon were the major depositories for ceramics used as burial goods following local beliefs. From the 15th century onwards, several shipwrecks such as *Pandanan*, found near southern Palawan, *Lena Shoal* sunk offshore of Busuanga, north of Sulu, and *Santa Cruz* in southern Luzon. All these ships were loaded with Chinese and other ceramics from the Southeast Asian region, which indicates an active ceramic trade between the Philippines and China, Thailand, Vietnam, and Burma from the mid to late 15th century. Most of the shipwrecks found in the Philippines from this period are concentrated in the southern part of Sulu, and many Southeast Asian ceramics have been found together with Chinese ceramics. In the case of the *Pandanan* wreck, out of 4,722 pieces recovered, 97% were Cham wares, meaning that trade in this region was geared towards southeast-Asian waters and was strongly connected with mid-southern Vietnam, Thai, and Borneo.

After the commencement of the Manila galleon trade, commerce between Manila and Macao became important and Macao especially became the conduit for the introduction of Chinese ceramics to Manila from Guangdong. Other Chinese pieces, especially those produced in the Zhangzhou area, were possibly brought by the *sangleyes* coming from Fujian. That is to say that between China and Manila there were at least two or three routes along which Chinese ceramics might be imported. One was, of course, to acquire them from the Portuguese merchants of Macao or from the large ports of Fujian province. According to Lucille Chia, Chinese junks coming from southern China to the Philippines, especially in the 17th century, were from Zhangzhou and Yuegang (月港), and Xiamen (厦門). The traveller Francisco Carletti confirms that porcelain was sold in the Guangdong market.

Recent studies on shipwrecks in the South China Sea, as well as land sites in Macao, Hong Kong, and Manila reveal interesting flows of Chinese ceramics in Asia during the 16th and 17th centuries. These include:

1. The *San Isidro*, which was found on the coast north of Manila and dated to the latter half of the 16th century. Almost all the ceramics recovered were early exports from Zhangzhou. These finds are evidence that ships from Fujian ports were coming in directly to Luzon via another supplier of ceramics other than the Portuguese.
2. Chinese junk *Binh Thuan* found off southern Vietnam and dated 1620-1640s. The ship was loaded with Zhangzhou wares and was probably distributing these ceramics to several ports in Southeast Asia. Considering the quantity of Zhangzhou wares loaded in this junk, the probable despatch port was Quangzhou, Amoy, or Zhangzhou in Fujian, or Shantou in Guangdong province. Considering the location, the destination was possibly the Malay Peninsula, Siam or Java.

Important land sites which hold the key to understanding the ceramic flows between Guangdong, Macao and Manila include:

1. The Macao city finds from Monte Fortess.
2. The Shangchuan Island (上川島) excavations.
3. The Ayuntamiento excavations.

According to Takenori Nogami an excavation of the Monte Fortes (built in 1622) was carried out in Macao City, and excavated ceramics are stored in the Macao

Museum of Art. Since Japanese *Hizen* (肥前) wares were the focus of this study, further research on the Chinese ceramics are necessary. The Shangchuan excavations, carried out in 2002, shows the importance of island to the Portuguese and Chinese trade going west. The ceramic finds from this site were mostly Jingdezhen wares of blue-and-white, and some enamel glazed wares, and mostly date between the reigns of Zhengde正德(1502-1521) and Jiajing嘉靖(1522-1566).
This is because the Ming emperor did not permit the Portuguese to settle in continental China, and during this period they were clandestinely trading with Chinese merchants on this island.

Among the pieces exported to Manila some were then transhipped to New Spain, although some stayed in Manila or other places in the Philippine islands. Excavated pieces from the *Intramuros* (presently called the *Ayuntamiento* site) were studied by archaeologists and they reported that among the ceramic finds 90% were Chinese ceramics dated to the late 16th to early 17th century.
These pieces are mass produced wares which can also be found in Mexico City in abundance. Some types are identical to those found in the *San Diego* wreck and are thus good examples of Chinese ceramics distributed and consumed by the Spanish in the Philippines.

2. Chronology and typology of exported Chinese porcelain excavated in the city of Mexico and its change during the 16th and 17th centuries

The excavation of the Zócalo area in Mexico City began in the 1970s with the city's expansion. The zone includes the historical area of the city of Mexico, where the earliest buildings of the colonial period were constructed from right after the conquest by Hernan Cortes, and functioned as the centre of the whole viceroyalty. Even today the area plays a major role in city life. The cathedral, presidential palace, city hall, main plaza, mint, archbishop's residence, the Franciscan monastery, and many other important buildings and residences of the first conquistadores and later wealthy Spanish were all concentrated in this area.

Naturally, due to this area's historical background, a large quantity of the Chinese ceramics exported to New Spain was destined to be consumed here. The author has carried out a thorough study of ceramic finds from the sites of Donceles Street, the Templo Mayor, the Cathedral, the Justo Sierra and the Santa Teresa. The Donceles Street site was a residential plot that has seen many changes over the course of time and it was difficult to chart how the land use was reorganised (see Map 2).

On the other hand, the Templo Mayor site is the former principal temple of the city of Tenochtitlan, where religious ceremonies took place before the Spanish conquest. What the building was converted into during the colonial

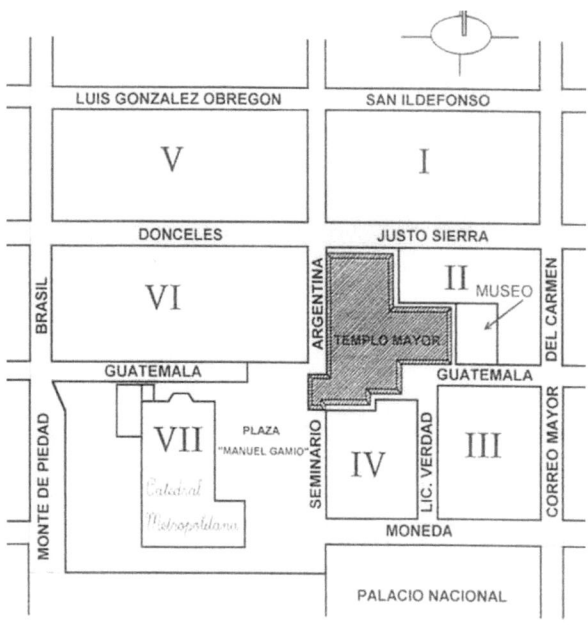

MAP 2: MAP OF EXCAVATED AREA OF ZÓCALO (©INAH).

period is not clear, although it is adjacent to the Franciscan monastery. Jingdezhen blue-and-white wares from the late 16th to early 17th century are the most abundant finds. 1,768 sherds were located by the present author in the storage room of the museum between 2007 and 2009.
Other provenances include Zhangzhou (漳洲) blue-and-white wares, Dehua德化 (Fujian) blue-and-white wares, Japanese Hizen (肥前) blue- and-white wares, and kilns from Guangdong. Another site, Santa Teresa, an ex-convent next to the Templo Mayor is another excavation area where 831 fragments were found. The origins of the fragments were mostly Jingdezhen, followed by Zhangzhou, Dehua, and some provincial kilns of Guangdong (see Figure 4).

The finds from these excavated sites can be classified into the following five phases.

1. Mid 16th century to 1575.
2. 1575 to the early 17th century.
3. First half of the 17th century.
4. Recession period: mid 17th century.
5. The end of the 17th to the 18th century.

Phase 1 includes pieces similar to those of some of the heirloom pieces in Portugal and dated to the Jiajing (嘉靖) period. Only six finds were made from the Mexico City, indicating the modest levels of ceramic trade during this phase. The archaeological context for dating these finds is related to the *Fort San Sebastien* wreck, a Portuguese ship which was lost off Mozambique between 1553 and 1558. The ship was loaded with pepper, nutmeg, mace and Chinese porcelain. Large deep bowls with flat rims, other bowls, large dishes and plates were found. Most of the motifs are of Jiajing style transitioning towards

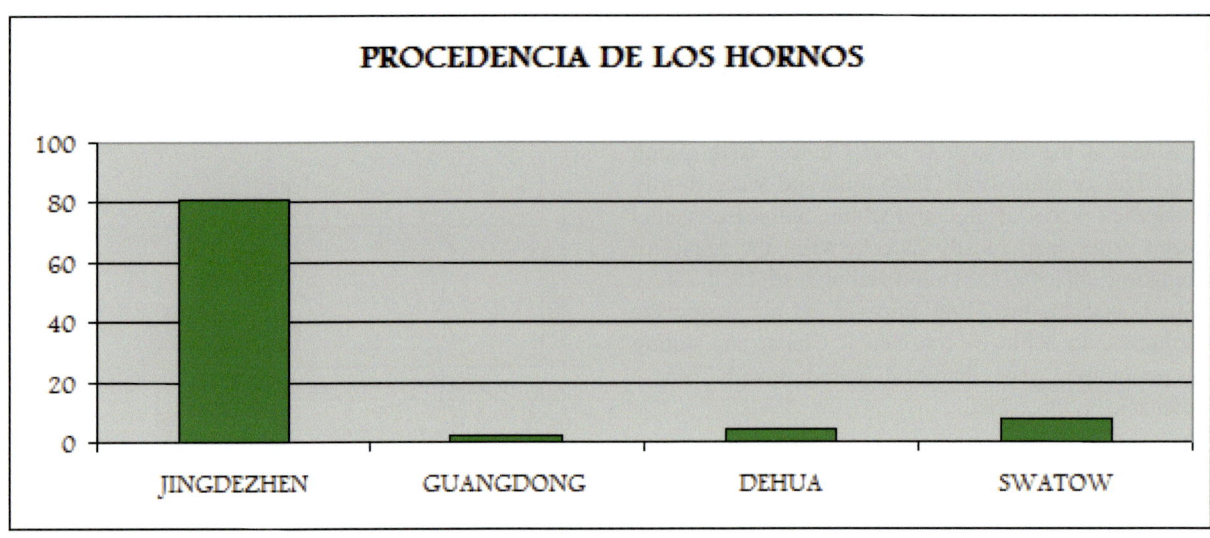

FIGURE 4: ORIGINAL PRODUCTION SITES OF CERAMICS FOUND IN MEXICO CITY.

PHOTOGRAPH 7: *KRAAK* PORCELAIN PLATE RECOVERED FROM THE *WITTE LEEUW* © RIJKSMUSEUM

Wanli. There are no *kraak*-type finds, although some flat rims are decorated with Taoist ribboned emblems and the peach sprays which became popular motifs, especially from the third quarter of the 16th century. Thus, the dating of this wreck derives from the matches with the document and predates the early Manila galleon trade and the porcelain finds can help us understand the early ceramic finds of Zócalo.

Phase 2 can be referred to as the most prosperous period of the Manila galleon trade. The finds include many plates and bowls of blue-and-white *kraak*-type wares which were popular in the American and European markets, and thus frequently included in ceramic exports to Portugal, America, and the Netherlands. It can be difficult to accurately define kraak ware, but here it can be generally referred to as plates, bowls, jars and 'kendis' with vertical divisions of panels with floral sprays, Taoist symbols and bamboo designs drawn on the cavetto. The central medallion is decorated with a Taoist symbol, deer and *taihu* (太湖) rockery, insects, birds, etc. (see Photograph 7).

It is true that *kraak* wares had a large market in Europe and America, as previously mentioned, although it was also popular in China and other Asian regions. There are four sites that provide the criteria for dating this period:

1. Ichijodani (一乗谷): A castle site located in Fukui Prefecture (福井県), Japan. This location was excavated in the 1980s. It featues a castle founded in 1471 and burnt down in 1554, thus the excavated ceramics can be dated to within around 80 years. Masatoshi Ono has analysed the finds and established a detailed chronology of the blue- and- white wares from this period.

2. Drake's Bay (1579-1595): This ceramic assemblage was found in Drake's Bay in California and dates between 1575 and 1595. One group of fragments were from the cargo of a Spanish galleon from central America raided by the famous English pirate Francis Drake in 1579. He brought the cargo to the California coast in order to repair his ship, the *Golden Hind*. He left the porcelain cargo or traded it with locals there (Coastal Miwoks) and set sail to cross the Pacific. Another cargo came from the *San Agustin*, a Spanish galleon commanded by Sebastian Rodriguez Ceremeño and wrecked off this coast in 1595. A study of this material was undertaken by George Kuwayama, based on field researches carried out by the University of California, Berkley, in the 1940s, and succeeded by archaeologists from San Francisco State University in the 1960s and 1970s. he Kuwayama concluded that the ceramic sherds recovered from the coast can be classified into two groups. The first group consists of sherds not eroded by sea

water and thus associated with the *Golden Hind* material deposited ashore. These are mostly Jingdezhen blue-and-white, early *kraak*-type wares, including plates moulded into ten bracket lobes with ten floral sprays decorated near the rim. The second group consists of sherds that show signs of having been eroded by water, and includes Fujian blue-and-white ware, a type totally absent from the sherds of the *Golden Hind*.

3. *San Diego* (1600): This Spanish galleon was sunk by the Dutch fleet in 1600 and found near Fortune Island, off the west coast of Luzon. The ship contained large amounts of Chinese blue-and-white porcelain from Jingdezhen, including many *kraak*-type plates with deer and *taihu* rockery design on the centre medallion. Other *kraak* wares are bowls with deer designs, large jars with covers with *ruyi* (如意) motifs on the shoulders and hermits drawn on each of the window panels of the body. Thai storage jars, Burmese 'Martaban' jars, and Chinese jars from Fujian or Guangdong were also found in this ship.

4. *Nossa Senhora dos Mártires* (1608): A Portuguese ship, loaded with pepper and Chinese and other Southeast Asian ceramics, left Cochin in 1606 and was lost the same year in the Bay of Cascais. Among the finds were examples of so-called 'Tradescant Jars', probably produced somewhere in southern China and featuring green and yellow glazes with designs of *qilin* (麒麟) and dragons. Other finds include Martaban jars and large Chinese brown-glazed jars. *Kraak* porcelain of good quality is abundant among all the ceramic finds. The decorative styles are more or less identical to those featured on the ceramics from the above-mentioned *San Diego* wreck. Many kraak-type plates designed with deer and *taihu* rockery are also present in both wrecks.

This phase is referred to as being a prosperous period for the town of Jingdezhen. During the Wanli period it is noted that around 20,000 craftsmen were engaged in ceramic production and the town was considered as the richest in the whole province of Jiangxi (江西省). In terms of the distribution of these ceramics, Shinan (新安) and Shanxi (山西) merchants were known to be active, especially the traders from Shinan, who came from the adjacent province and so had a geographical advantage. They may also have been in some way involved in the exports of Chinese wares to the West, via Guangdong or Fujian merchants, some of them receiving special requests from religious orders and other elites.

In **Phase 3** the development of the Pacific trade loses some of its momentum in the mid 17th century, when China experiences civil war and all trade activities in the Chinese coastal area are prohibited. The contexts for dating this phase include the following:

1. The *Witte Leeuw* cargoes (1613): This Dutch ship set sail from Banten and was lost near St. Helena in 1613. Cargo finds from this ship include pepper, nutmeg, guns, and Chinese and European ceramics. 300 complete or reconstituted Chinese porcelains were taken from this site, along with 300-400kg of ceramic sherds. The majority of the Chinese ceramics consisted of high-quality, *kraak*-ware dishes. All of these *kraak* finds are Jingdezhen blue- and-white wares, moulded in eight batches or more. Peach sprays or Taoist symbols are drawn in each of the panels, and between the panels there are vertical belts with bead decoration. The central medallions are painted with flower-baskets, landscapes, birds and Taoist symbols.

    These *kraak* plates from the *Witte Leeuw* wreck are stylistically more varied in decoration, using brighter and denser cobalt than those finds from the *San Diego* wreck.

2. The *Hatcher* cargo (1643-1646): This was a Chinese junk loaded with ceramics for export and sunk in the South China Sea around 1640. The junk was salvaged by Michael Hatcher and 23,000 porcelain ceramics were recovered for sale by the auctioneers Christie's. The junk is thought to have made regular crossings between Fuzhou, Zhangzhou, Ninpo, Taiwan, and Batavia.

    Save for a few large Zhangzhou glazed plates, most of the pieces salvaged were blue-and-white Jingdezhen porcelain, although these were especially selected items from the seabed and there must have been other jars and ceramics underneath. Some of the *kraak* wares display stylistic deterioration, with some of the brushwork being rather rough when compared to the pieces from the *Witte Leeuw* cargo. Irrespective of the large quantity of *kraak*-type materials from the *Hatcher* cargo, the production of this type of ceramics must have been entering its final decades. However, among the remarkable pieces, which are only found in this cargo, are some covered jars with large floral motifs and designs of phoenix, dragon and *qilin* daringly drawn on the body. These are carefully composed pieces compared to the other *kraak* wares present, showing the growing popularity of other forms destined for European markets. Some of these new styles include new forms for export, such as tall vases, and large covered oval jars, probably intended for in house ornaments in Europe. The *Hatcher* cargo pieces show us a change in Jingdezhen production, which probably derived from the rapid growth of European markets, and even though the late Ming civil war might have influenced official trade and kiln productions, private trade was still active and kiln production was still functioning well enough to meet European demands.

Phase 4 marks a period of recession, from approximately 1650 to 1690, during which no fragments could be identified in correlation with any wreck or land sites. The fact that mid-17th century pieces cannot be found possibly reflects the historical fact that China was then in a state of civil war, and from the late Ming and early Qing dynasties trade was strictly banned along the coast. Jingdezhen was also in turmoil, and in 1674 Jiangxi province was controlled by an army led by Wu Sangui (吳三桂), a Chinese military general who played an important role in the late Ming period and the establishment of the Qing dynasty. It is said that half of Jingdezhen town was destroyed by fire only 20-30% of its population was therefore able to work in ceramic production.

Phase 5 coincides with the late 17th century. At least in New Spain, the ceramic types from the latter half of the 17th century (corresponding to the 1690s) are identical to finds from the *Vung Tau* wreck. This sunken vessel was discovered by fishermen off the islands of Con Dao, southern Vietnam, and a salvage company recovered 48,000 ceramic items from it in 1991. The ship was said to have set sail from China, heading for Batavia. The cargo was full of blue-and-white cups and saucers for tea drinking, tall vases, and some large Zhangzhou plates. Small cups and saucers for tea and coffee, as well as tall cups for chocolate are also known from this period, thus indicating the popularity of tea, coffee, and chocolate drinking in Europe and America.

Phase 1: Mid 16th to 1575

Type A

At least 12 pieces that fall into this period are known from Zócalo. One is a large dish with central medallion surrounded by *ruyi* head and central vegetal design (see Photograph 7 and 8).

Type B

A large, deep dish with floral scrolls drawn on the exterior and with central scrolls and other motifs (Photograph 10). The outlines are neatly drawn and coloured inside with light cobalt.

Type C

A dish with flat rim with scroll design. Cavettos and central parts are in many pieces and are left blank or sometimes illustrated with scrolls and egrets (Photograph 11). Complete pieces of these sherds are found among heirloom collections in Portugal and are dated to around the mid 16th century. These are probably the earliest types that were exported from Manila to Acapulco. There is a correlation between heirloom items and some other finds from the *Fort San Sebastian* wreck in chronological terms. In addition, the remark in the above-mentioned document that in the early galleon trade there were some Portuguese merchants sending Chinese ceramics to New Spain suggests that these were specially produced items for the Portuguese market and were probably brought to Manila by the Portuguese, who already greatly appreciated Chinese porcelain.

Phase 2: 1575 to the early 17th century

Most of the excavated ceramics from Zócalo fall into this period. From the Templo Mayor site in particular the majority of finds was produced from c. 1575 to the early 17th century.

Type D

There are several ceramic styles from this period, including the bowl type that is generally referred to by the Japanese archaeologist as *wantouxin* (万頭心). These are relatively small bowls with a slightly raised base and decorated with a crab, hermit and central peach-flower motifs (see Photograph 12). This is a type commonly found at the *Ichijodani* site, and, based on the stratum, Masatoshi Ono has dated it from the third quarter, to the end of the 16th century.

Type E

Features a plate with standing phoenix on the central medallion and Taoist symbols on the rim. This type is know from many other sites from Lisbon and northwestern Spain to southeastern Asia. The exact same type has not been found on any datable wreck of this period, although stylistically, and observing the deterioration of the rim decoration and drawing, and rough manufacture compared to that of mid-16th century pieces, it can be dated to 1570s, towards the end of the century just before the *kraak* type came into mass production. It is also dated by Masatoshi Ono from around 1570 to the early 17th century, although these pieces are absent from both the *San Diego* and *Nossa Senhora dos Mártires* wrecks (see Photograph 13).

PHOTOGRAPH 8: JINGDEZHEN BLUE-AND-WHITE SHERD EXCAVATED FROM THE TEMPLO MAYOR, MEXICO CITY © INAH.

Chapter III - Exported Chinese Porcelain in New Spain

Photograph 9: Jingdezhen blue and white plate © Fundação Almeida

Photograph 10: Jingdezhen blue-and-white sherds with vegetal motif excavated from the Donceles Street site, Mexico City (© INAH).

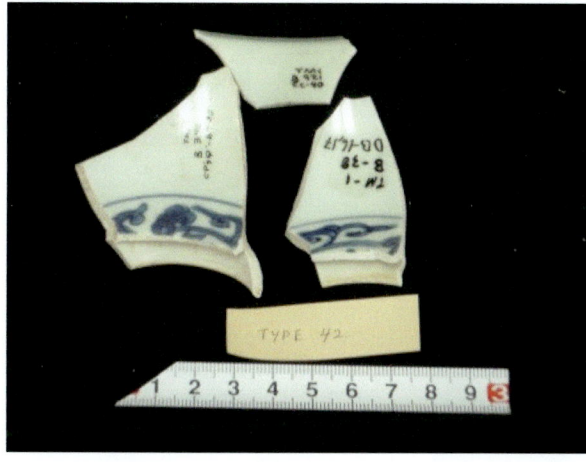
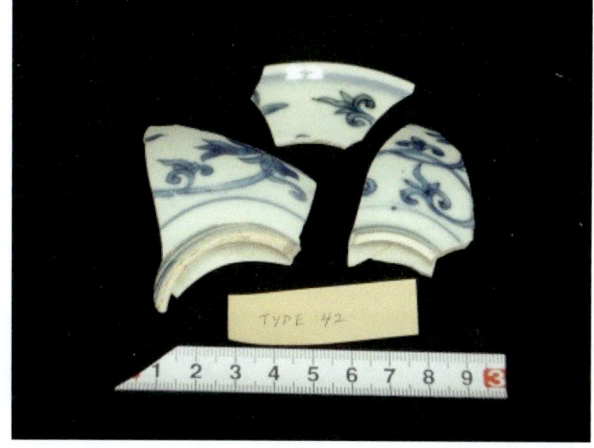

Photograph 11: Jingdezhen blue-and-white sherds excavated from the Templo Mayor, Mexico City (© INAH).

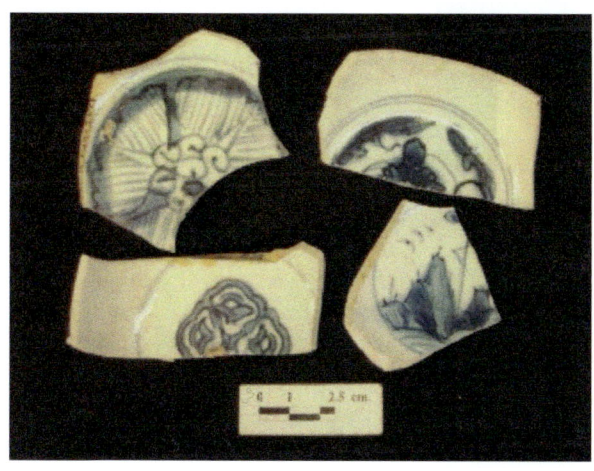
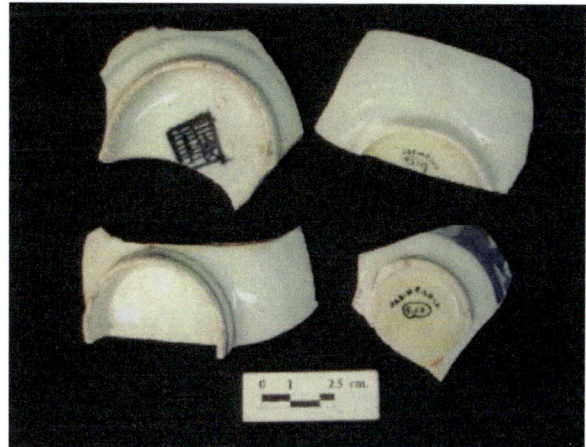

Photograph 12: Jingdezhen blue-and-white sherds excavated from the Templo Mayor, Mexico City (© INAH).

Type F

Another popular type from this period are the early *kraak* wares, which coincide with the *Golden Hind* pieces. These pieces are drawn with sensitive brushwork in greyish light blue. Many pieces of this type have two deers with scenes of *taihu* rockery, *ruyi* clouds and pine trees drawn in the centre and the cavetto is divided into

PHOTOGRAPH 13: JINGDEZHEN BLUE-AND-WHITE PLATE WITH PHOENIX MOTIF EXCAVATED FROM THE TEMPLO MAYOR, MEXICO CITY (© INAH).

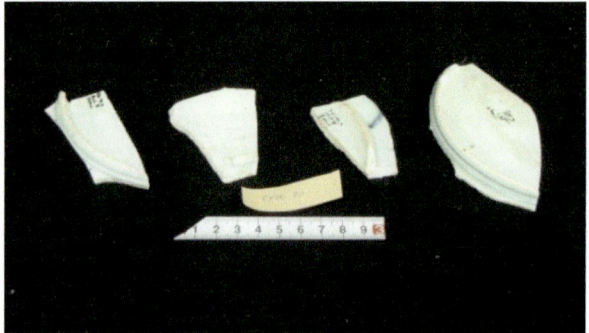

PHOTOGRAPH 14: JINGDEZHEN BLUE-AND-WHITE SHERDS OF PLATES FROM THE TEMPLO MAYOR, MEXICO CITY (© INAH).

PHOTOGRAPH 15: JINGDEZHEN BLUE AND WHITE PLATE FUNDAÇÃO ANASTÁSIO GONÇALVES

eight to ten panels with each panel decorated with floral spray, insects, and Taoist symbols with beads and tassels in between. These are also present among the *San Diego* and *Nossa Senhora dos Mártires* finds were probably popular for some decades, from around 1579 to 1610 at the latest (Photographs 14, 15 and 16).

These probably form the earliest type of *kraak* wares, considering the dating of the *Golden Hind*, and were produced in abundance until the early 17th century. New variations of *kraak* wares are also found on this wreck site, but at this time (1608) this earliest type was already in decline.

Type G

Another type known from this period is the *kraak* bowl from *San Diego* with vertically divided panels and drawn Taoist symbols with floral and bamboo motifs (Photograph 17).

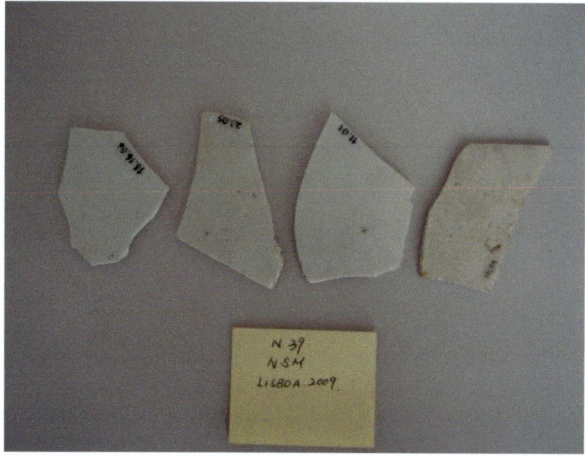

PHOTOGRAPH 16: JINGDEZHEN BLUE-AND-WHITE SHERDS FROM *NOSSA SENHORA DOS MÁRITIRES* (©CNAS-DGPC).

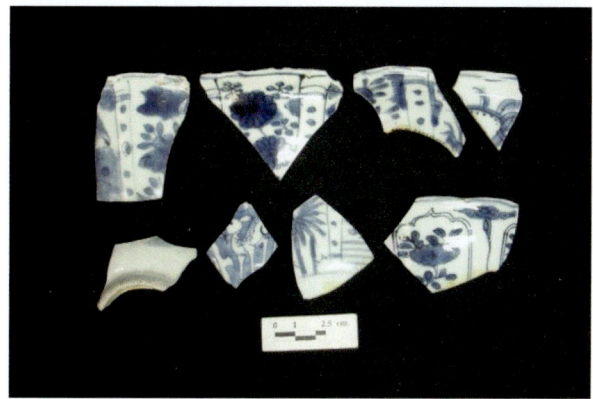
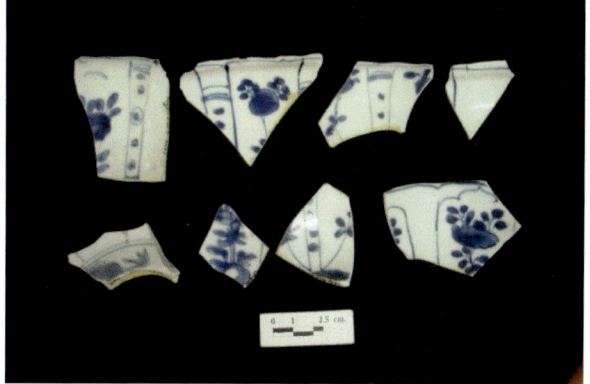

PHOTOGRAPH 17: JINGDEZHEN BLUE AND WHITE SHERDS OF *KRAAK* BOWLS EXCAVATED FROM THE TEMPLO MAYOR, MEXICO CITY (© INAH).

PHOTOGRAPH 18: JINGDEZHEN BLUE AND WHITE SHERD EXCAVATED FROM THE TEMPLO MAYOR, MEXICO CITY (© INAH).

Type H

A bowl with 'flying horse' motif and tassel design on the exterior body can also be found among other ceramics. The interior of the sherd found from the Templo Mayor is decorated with slightly incised flowers in a technique called *anhua* (暗花) (Photograph 18).

Phase 3. First half of the 17th century

Type I

There is a group of well-drawn large dishes of *kraak* type from this period that match finds from the *Witte Leeuw* cargo. The central medallion is described with Taoist

 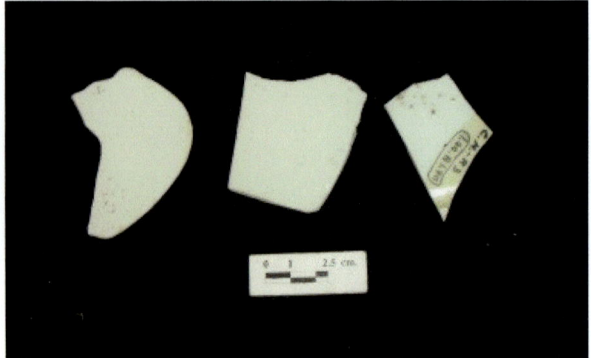

PHOTOGRAPH 19: JINGDEZHEN BLUE-AND-WHITE SHERDS EXCAVATED FROM THE CATHEDRAL SITE (ZÓCALO, MEXICO CITY-© INAH).

PHOTOGRAPH 20: JINGDEZHEN BLUE-AND-WHITE SHERD EXCAVATED FROM ZÓCALO, MEXICO CITY (© INAH).

symbols and flower-baskets, while the cavetto is neatly decorated using dark cobalt (Photograph 19).

Type J

A plate of *kraak* type generally called *meizande* (明山手) with the panel divisions described in small medallions. The centre is decorated with a horse design, although there are several decorations featuring central medallions with Taoist symbols, the grasshopper and flower combination, birds, ducks and frogs. This *kraak* type probably came into production after 1600 since they are absent from the *San Diego* wreck. On the other hand these plates are abundantly found among the *Witte Leeuw* cargoes (Photograph 20).

Type K

Represents a group of *kraak* plates that are roughly drawn and have more unpainted areas on them than the *kraak* plates from the *Witte Leeuw* cargoes. These plates mark the decline of *kraak* wares and can also be found among pieces from the *Hatcher* cargo, some three decades later than the *Witte Leeuw kraak* wares (Photograph 21).

Type L

Consists of a group of plates roughly drawn with a central deer motif and a simple *kraak* cavetto with beads and floral spray. These remains have been found at both the Templo Mayor and the Donceles Street sites in Mexico City. A feature is that the cobalt is darker and more tinted than other ceramics. Although it still maintains the form of *kraak* decoration, with divisions and central medallion, the brushwork is extremely rough. Particles of sand were found adhered to the foot ring. This type was also found in the *Hatcher* cargo and thus can be dated to first half of the 17th century (Photograph 22).

Phase 4. Mid 17th century

So far there has only been one piece found which falls into this period; it is generally referred to as a 'crow cup' from its central bird motif. There are also other transitional pieces from heirlooms from this period. The extreme scarcity of finds dated to this period shows a recession in terms of the ceramic trade between Manila and Acapulco (Photograph 23).

CHAPTER III - EXPORTED CHINESE PORCELAIN IN NEW SPAIN

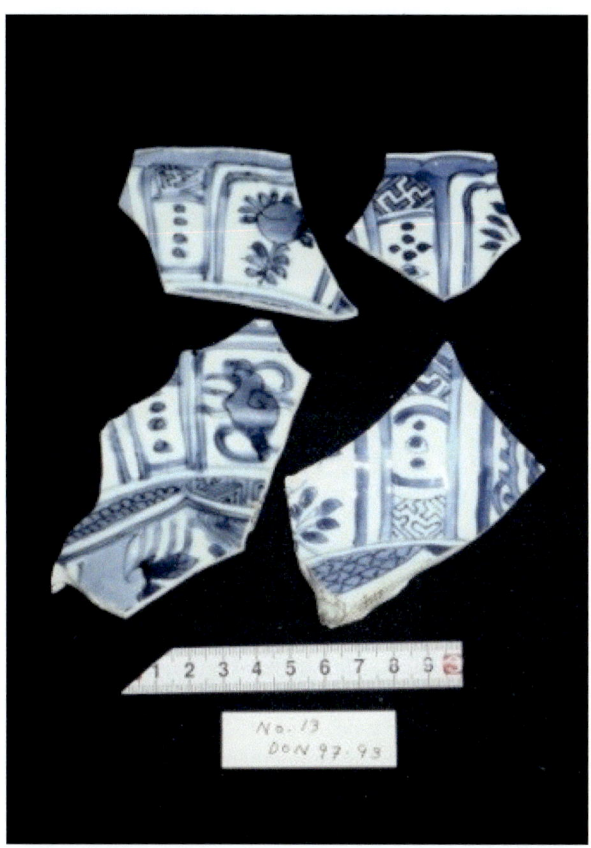 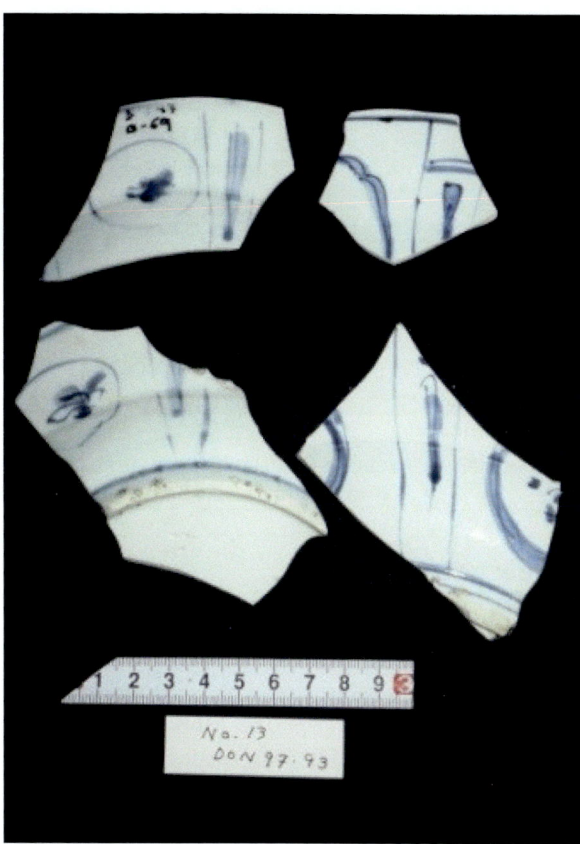

PHOTOGRAPH 21: JINGDEZHEN BLUE AND WHITE SHERD EXCAVATED FROM THE DONCELES STREET EXCAVATIONS, MEXICO CITY (© INAH).

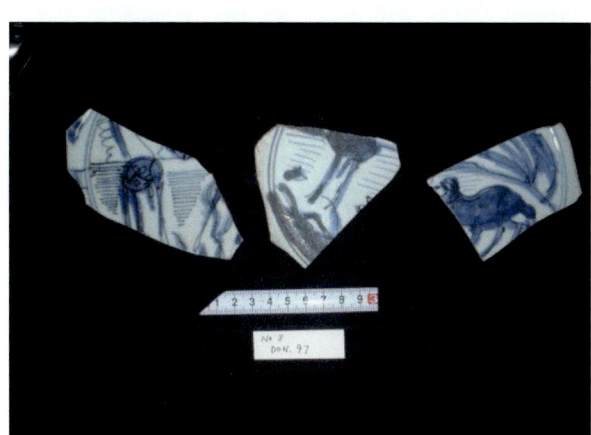 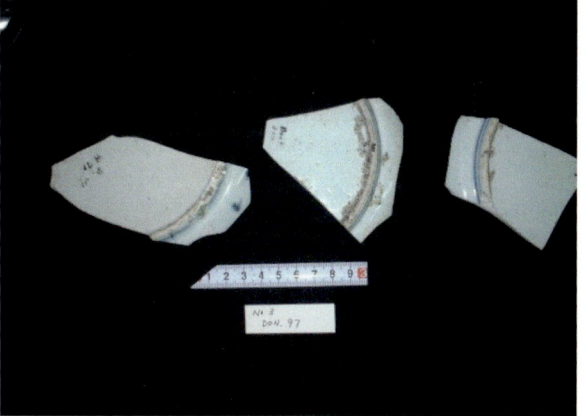

PHOTOGRAPH 22: JINGDEZHEN BLUE AND WHITE SHERDS EXCAVATED FROM THE DONCELES STREET EXCAVATIONS, MEXICO CITY (© INAH).

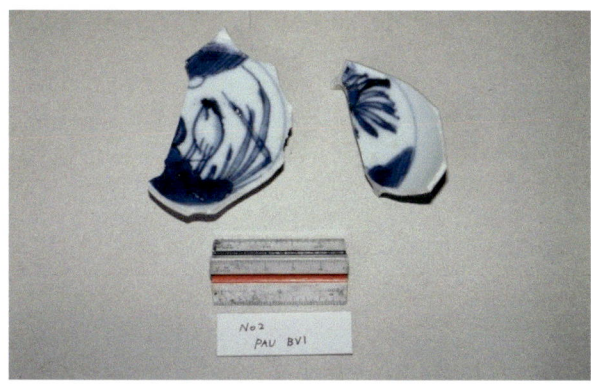 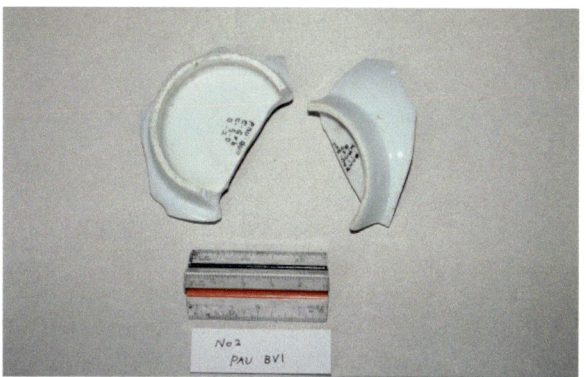

PHOTOGRAPH 23: JINGDEZHEN BLUE AND WHITE SHERDS OF "CROW CUP" EXCAVATED FROM ZÓCALO, MEXICO CITY (© INAH).

PHOTOGRAPH 24: JINGDEZHEN BLUE-AND-WHITE SHERDS EXCAVATED FROM THE TEMPLO MAYOR, MEXICO CITY (LEFT). JINGDEZHEN BLUE AND WHITE SHERDS EXCAVATED FROM LA CALLE LICENCIADO VERDAD (RIGHT) (© INAH).

Phase 5. 1690s onwards

Type M

Some of the finds correspond with what are generally known as *Kangxi* (康熙) pieces. Most of the remains from this period are tall cups; plates, which in the earlier period represented the main form, are now rarer (Photograph 24). This coincides with the craze for drinking chocolate, especially among the wealthier classes. Although there has been no specific study on chocolate cups in New Spain, it may be considered that they were used for drinking both coffee and chocolate. In other cases, especially in Europe, these cups were sometimes used as chalices, with silver guilt, in the Catholic mass, although in Mexico City these are found in abundance and were probably for daily use.

Overall there were 11 types classified among the excavated pieces from the Zócalo area in Mexico City, the major part being from the Templo Mayor and Donceles Street sites. Analysing the fragments found from the Templo Mayor, the results show that 74% fall into Phase 2 (1575 to the early 17th century). This matches the high percentage of fragments from this period found at the Ayuntamiento site in Manila (see Figure 4).

Finds of the early type A (c. 1550 to 1575) are few, although it should be remembered that in 1565 the first galleon was despatched from Cebu and from then on trade grew rapidly towards the third quarter of that century. It is also important to note that the earliest pieces were identical with the heirloom types found in Portugal, rather than the styles more commonly found in Asia. This may be suggest that from the beginning of the Manila galleon trade there were interventions by the Portuguese, who knew the European market for Chinese ceramics and had direct access to China via Macao.

The abundance of Phase 2 finds indicates the link in the increase of the Manila galleon trade and the growth of Jingdezhen production. The Ming ban was officially abolished in the mid 16th century and began a policy of mass exports to the West via Macao. From the

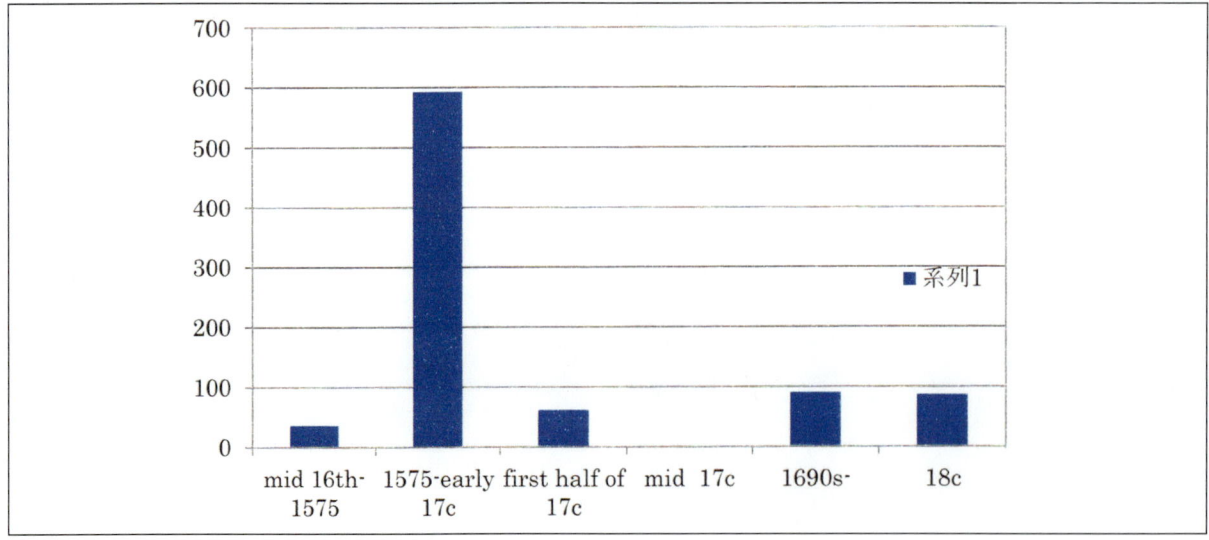

FIGURE 5: CHRONOLOGICAL DATA BASED ON THE CHINESE CERAMICS FOUND AT THE Zócalo área, MEXICO CITY.

beginning of the 16th century the main entranceway to European markets for Chinese ceramics was Lisbon and distribution increased rapidly after this time.

The Spanish relied on Portuguese and the Chinese merchants for the acquisition of Chinese products. Analysis of Chinese ceramics excavated in New Spain provides clues that the trade tie between Macao–Manila–Acapulco may have been stronger than between China and Manila. In other words there were different roles for Portuguese and Chinese merchants coming to Manila. The *San Isidro* wreck, loaded with Zhangzhou wares, is a good example. Portuguese merchants supplied Jingdezhen porcelain for re-export to New Spain and to be consumed by the richer merchants, while Chinese merchants, primarily from Fujian, supplied Zhangzhou wares mainly to local markets. Ceramic types found in the Mexico City clearly match with Portuguese heirloom finds and recovered pieces from Portuguese wrecks, and are forms which would adapt well to western tastes, such as flat plates and cups. Although it is difficult to ascertain the precise numbers of ceramics exported to New Spain, there is a primary source which may help greatly. In 1640, 13 large wicker baskets containing Chinese ceramics entered Mexico City and some of these were subsequently registered (see Appendix). Exactly how many items were packed in each basket is unknown, but it may have been similar to the quantities found in Dutch consignments. In any event it does indicate an outline of ceramic flows in selected documents, together with other Asian products (Photograph 25).

The gradual decline of the porcelain trade towards the middle of the 17th century may be explained in relation to the Inquisition in Mexico and the persecution of some wealthy *converso* merchants, who fled Mexico, and causing Spain to lose its strong tie with Macao. It is highly probable that the conflict with the Dutch in southeast and east Asia hastenend the gradual decline in trade with China during this period. In addition, the extreme scarcity of Chinese ceramics from the middle to the end of the 17th century can be explained by the civil wars of the late Ming and early Qing dynasties, leading to the 'Great Evacuation' of the Qing dynasty to expel the warlord Koxinga (鄭成功) who was gaining great wealth from trade around the South China Sea. The reappearance of Chinese ceramics in Mexico from around 1690 corresponds with the end of the coastal evacuation in 1683 and the subsequent end of the depression in the Chinese economy brought about by the firm leadership of the Qing dynasty. The active trade carried out by the Fujianese merchants from c. 1680 was maintained until c. 1740. Ships from Macao still sailed to Manila although the traffic markedly decreased.

3. Material culture and porcelain in the society of New Spain

As mentioned above, there were regular flows of exports of Chinese ceramics to New Spain from the mid 16th and into the 17th century, except for an hiatus from c. 1650 to 1690. From practically all of the excavation sites in the Zócalo area of Mexico City Chinese ceramics have been recovered (i.e. the Templo Mayor, Donceles Street, the cathedral, Justo Sierra, Santa Teresa, La Calle de Licenciado Verdad and Casa de Moneda). Although the total numbers of Chinese ceramics found from all the excavation sites cannot be precisely counted, it is easily imagined that very large quantities were traded and brought to New Spain.

The spatial distribution of these ceramics, which is concentrated in the central part of Mexico City, indicates that they were much appreciated items by those classes of society permitted to live within the city. Appreciation of Chinese ceramics among the wealthy is also substantiated by the numbers of heirloom pieces (Photograph 26).

These Chinese ceramics, especially imports from Jingdezhen, were therefore highly prized for their elegant forms, glazes and sophisticated decorations, which had never been seen in the Americas before. The impact on the material culture of New Spain must have been great and influences can be seen in many examples of Talavera wares from Puebla.

When the Spanish conquered Mexico there was already a highly developed local ceramic production, although the Aztecs did not seem to know the wheel, glazes, or kiln firing. Such were the differences between the ceramics of New Spain and Metrópoli that the Spanish soon began to import their traditional ceramics. Ceramics from the Iberian Peninsula introduced many

PHOTOGRAPH 25: WICKER BASKET CONTAINING JAPANESE CERAMICS (©AMSTERDAM CITY MUSEUM)

Photograph 26: Gilded Jingdezhen polychrome vase decorated in ex-private residence in Zócalo, currently a restaurant (©INAH)

new forms, such as cooking pots, basins, jars, plates, bowls – indeed all the basic items in daily use in Iberian culture. The everyday needs of the Spanish residents in New Spain began to open up the market in ceramics, and consequently cultural exchanges between Spain and New Spain produced an early impact on Mexican ceramics. With these imports new technologies were introduced by the invaders, especially from Seville, where ceramic production was traditionally active. Local ceramic production gradually began to be substituted for imported Spanish ceramics in New Spain. In 1531 ceramics and tile kilns were established in the city of Puebla (Province of Los Angeles), which went on to produce the famous Talavera products and the leading city in New Spain for tile outputs all through the colonial period.

Chapter IV

Distribution of Chinese Ceramics and Asian Products in Spanish Society

After trade started between Manila and Acapulco, and a certain amount of Chinese ceramics began to arrive in New Spain, there was an escalation of ceramic production of Talavera wares in Puebla. Blue-and-white imitations of *kraak* types and decorations can be seen in several collections. Chinese motifs, such as insects, birds and floral sprays began to appear on Talavera wares, indicating the impact of Chinese ceramics generally. These local blue-and-white ceramics were probably cheaper, and, due to their technical quality, were considered to be inferior to Chinese ceramics. Even so, the demand for the Chinese taste in ceramics was great, and in order to meet this Talavera wares were influenced by Oriental decoration. Consequently, this Chinese style was applied to many other sources of ceramic production – such as Peru, Panama, and Guatemala. This meant that almost all social classes could possess 'Chinese' porcelain and this indicates how imports of Chinese goods changed the look of ceramic decoration of the time in America, and even in Europe, and may be considered as part of the proliferation universally of the taste in Chinese ceramics. Talavera wares developed their own style, absorbing European cultural elements and flourishing as unique products in New Spain.

Chinese ceramics also underwent significant changes to adapt to European tastes and special orders, and gradually lost their 'Asian' elements in the 18th century and developed original designs for these new markets. Decorations were commissioned in Guangdong, on the interventions of agents, to satisfy the demands of the European and American markets. Thus, Chinese ceramics were culturally absorbed by European tastes, leading to the style known as 'Chinoisserie', combining the Chinese and European elements that went on to dominate Europe and European culture in America.

Spain materially developed its commerce with the New World, especially during the 16th and 17th centuries, and widened its markets, importing products such as silver, gold, indigo, hides and other goods. Quantities of Asian products entered through the port of Seville, although these never matched the amounts entering the Portuguese markets, where Asian goods were far more appreciated by society and large volumes of porcelain and other Asian goods are to be found among the excavated materials and heirloom pieces. The 16th century was an era when Portugal began to import Asian goods in large quantities for the first time in European history. Until then silks, spices, porcelain and other products were transported to the European world by Muslim merchants (indeed from the 10th century), although not in large quantities. Highly valued Chinese silks and ceramics (Yuan blue-and-white and celadon pieces) often appear in miniature paintings from the Middle East, and from there these luxuries were probably brought further west by Italian merchants. The sudden large import volumes of Chinese ceramics from southern China to Lisbon in the early 16th century indicate that there was a wide acceptance of Asian luxury goods in the country, based on earlier preferences for Asian goods. Imports to Lisbon were distributed to many Portuguese cities, such as Coimbra, Oporto, and further north towards the Galician coast (and possibly onwards to the Netherlands from the third quarter of the 16th century). The impact on European material culture of Asian luxury goods was not significant at this moment and 16th-century Asian products are scarce in European collections. However, it is clear that this distribution via Lisbon did stimulate the later demand for Asian goods, especially in the Dutch markets.

The Netherlands became active in Asian waters from the early 17th century and, indeed, often attacked other Europeans in the Molucas and the Philippines. The Dutch merchants were interested in importing spices, silks from Tonking, silver from Japan, porcelain and other goods. Chinese porcelain was largely imported to Amsterdam and distributed to the European markets.

Portugal was the key importer and distributor of Chinese ceramics throughout the 16th century, as previously mentioned, but the Netherlands took over this leadership and became the most important importer and distributor of Chinese ceramics in Europe in the 17th century. Much of the porcelain and other goods imported were redistributed to England, Germany, France and northern European countries. Large quantities of Asian goods was thus distributed within European markets via Amsterdam and had a great impact on the material culture of European society. Many elites bought porcelains and lacquer wares for their own collections and some of these in turn became important collections in later times.

Such was the acceptance of Asian goods in European society in the 16th and 17th centuries. However, silk products, porcelain and lacquer wares of this period in Spain are extremely rare in every collection and among the archaeological record. Nevertheless ceramic materials, which normally survive various harsh conditions, can be found in many modern city sites all over the world. In 16th- and 17th-century Spain, the excavated finds are few, even in an important port-city such as Seville. This in turn suggests that:

1. Asian goods were not generally accepted in Spanish culture and were thus not imported.
2. Chinese porcelain was not well recognized by archaeologists and thus does not appear in excavation reports.

The acceptance and distribution of Asian goods in Spain, especially during this period, has never been discussed in detail; whether they were 'present' or 'absent' as part of the material culture compared to other European societies, or 'how' they were accepted is not clear. It has been generally viewed that imports of Asian goods to Spain cannot be compared to the volumes imported by Portugal. However the true quantities of imported material have never been studied, and thus comparisons between the two societies are not open for academic discussion. The quantities and acceptance of Asian goods in Spanish society and culture are important for an understanding of the trends and character of the trade, as well as the actual trade routes followed.

The objective of this chapter is to clarify whether Asian goods (represented by Chinese porcelain) were imported and accepted by the Spanish population and its material culture. It is necessary to study the quantities of materials imported from archaeological sites and excavated pieces, and also to obtain information from the historical archives.

The research method focuses on the study of Chinese porcelain excavated in Seville, which was the major port of entry for Asian products coming from Veracruz, in order to examine the quantities and types of imported Chinese porcelain. The artefacts will be limited to excavated material from the historical area of the city, which is adjacent to the Guadalquivir River, where archaeological researches have been carried out within this decade. 38 fragments have been found so far and will be classified by their dating. An historical method will also be applied and documents will help in clarifying whether Asian products were actually imported via America and were brought into the port of Seville. The author has checked the official shipping record (*registro del barco*) in the *Archivo General de Indias* that dates from 1590 to 1600. This is an archive related to the inventory of all the cargoes loaded in galleons coming from America and entering the port of Seville. With this, the distribution of the Asian goods, especially focused on Chinese porcelain, will be discussed in order to clarify how they were accepted and treated in Spain during the 16th and 17th centuries.

1. Archival study of Asian products exported to Seville from Veracruz

To view the scale of trade and the nature of Asian products delivered to the port of Seville during this period, the present author has researched the registered cargo lists of ships entering the port from America between 1589 and 1600.

The inventories include products from Veracruz and other ports, such as Havana, and thus refer to many products from the Americas: silver, indigo and animal hides are the major products mentioned in this inventory, with Asian goods the minority.

The first register of Chinese goods appears in 1590 relating to a *flota* named the *Buen Jesús* that travelled from Veracruz to Seville with three boxes of Chinese ceramics, one of which contained '212 pieces of large and small ceramics from China'.

This box of ceramics was the personal property of the Archbishop of Mexico City, Pedro Moya de Contreras. In the next year, 1591, there is reference to Chinese products in terms of one box with two dozens Chinese ceramics, plates, bowls, ten pieces of Chinese ceramics and another box of Chinese ceramics. This was loaded on a galleon named the *San Bartolomé*, which left from the port of Veracruz. The latter box was sent from a man named Graviel de Balmaceda, a resident of Mexico City, to Lucas de Valorado in Seville as a 'present'. In the same year nine boxes containing various Asian products were transported from Veracruz to Seville on a galleon named the *Nuestra Señora de la Concepción*. Among these nine boxes, two were loaded with '10 arrobas and 14 libras' of silk. 1libra=1 pound.

Other goods were taffetas, damasks, silk clothes from Japan, other textiles, and curious religious artefacts such as an image of San José decorated with feathers. Another galleon named the *San Juan Bautista* carried 120 libras of silk and five boxes of Chinese products. These two items were sent by Gerónimo Perez Aparicio, a merchant residing in the city of Veracruz. This merchant's name appears three more times as having sent silk and ceramics until 1601 and therefore might have specialized in Asian products brought to Acapulco from Manila. The next reference to Asian goods can be found in a shipment sent in 1595. One box contained 98 pieces of large and small Chinese ceramics and another box had three dozen Chinese plates and bowls. This latter box were the personal belongings of a passenger named Alonso de Arroyo. A detailed and interesting reference to Chinese products appears in the same year as cargo in the galleon the *Nuestra Señora de Esperanza*. This galleon carried 3 boxes of ceramics costing 18 pesos each. In 1598 a galleon named the *San Francisco* set sail from Veracruz with a box of Chinese ceramics and textiles. The register refers to 'coloured plates', 'large bowls with gold decoration and coloured' and 'clay bowls from China'.

This is a rare reference to decorated Chinese ceramics. The items referred to as 'gold decorated and coloured large bowls' (*escudillas grandes doradas y coloradas*) are most probably Wanli (万歷) period bowls with polychrome decoration on the exterior. This type is coloured red and green with gold pigment painted over the exterior. The interiors feature peaches, crabs or floral motifs in cobalt in the centre. The type is found in excavations in Mexico and is usually dated to the last quarter of the 16th century. However the reference to clay bowls from China (*escudilla de barra de China*) probably indicates poorer quality stoneware items produced in southern China or southeastern Asia, although hardly any examples have been identified in Mexico to date.

Between 1590 and 1601, 488 pieces and four boxes of Chinese ceramics appear as cargo from New Spain to Seville. This is a very small quantity compared to silk products exported to Seville via Veracruz, which was 40 boxes in a decade. Other goods include various products referred to as *menudos* or 'cositas de China', of which there were 17 boxes. The quantity of ceramics traded to Seville is very small considering the number of ceramic cargoes loaded onto a single ship in other European countries. For instance a Dutch ship in the 17th century could take more than 2,000 pieces for a single merchant. The demand for Chinese ceramics in the Netherlands grew rapidly after the capture of two Portuguese *carracks* by the Dutch in 1602, and 1604 (the *Sanyago* and *Santa Catarina*, carrying more than 1,000 porcelain items each), and it is estimated that as many as 250,000 pieces of porcelain could be brought back to Europe in one vessel.

The ceramics transported to Seville are all 'plates' and 'bowls' and no other types are found – i.e. the jars that frequently appear in registers of trades from Manila to Mexico City. These jars from China and Southeast Asia were popular products for the Manila galleon trade as they could contain spices, such as peppers, and other items for consumption. This indicates that all the ceramics brought back to Seville were luxurious table wares and were limited to plates and bowls. Furthermore, some of the cargoes contained personal property, or personal presents, and were not intended for the regular markets where goods were sold and bought in shops.

2. Chinese porcelain excavated from Seville

The city of Seville served as the most important international and domestic port based on its geographical location. Using the Guadalquivir River, boats and ships were able to navigate into the city and the city was therefore already prosperous in Roman times and throughout the Middle Ages. After the discovery of the New World, the city became the only port serving the galleons from the Americas. All the products from the New World that entered this port were unloaded and redistributed within the country for the major interior cities and other port-cities in Europe. Seville was at this point the most active and prosperous city and it was where the wealth gained from trade with the Americas was concentrated. The luxury goods it handled were consumed by the elites in Madrid, although Seville also had many rich merchants wealthy enough to buy expensive merchandise and large quantities of luxury products therefore were destined for this city – silverwares, gold items, furniture, silks, and other goods. Chinese porcelain has been excavated at six sites in Seville, all inside the old city wall (see Map 3).

2-1. San Juan de Acre

The earliest structure dates back to the Muslim period of the 11th century that was later converted into a hospital (San Juan) in the 16th century. It existed into the 18th century. The site is located adjacent to the city wall that surrounds the entire historical part of the city and faces the Guadalquivir River. Chinese ceramics were excavated from the layer dating to the modern period associated with Lisboetan blue-and-white, various Italian ceramics, German and Belgium stoneware, Sevillana and Talavera wares.

2-2. Calle San Fernando

This is a site excavated during the construction of the metro line and the station of la Puerta de Jerez. An ancient wall of the *Almohade de Sevilla* (destroyed in the mid 19th century) was excavated. The original had six defence towers during the first half of the 12th and early 13th centuries. Within this site some remains from the 17th and 18th centuries, connected to the wall, were discovered, parts of which were constructed outside the 18-century tobacco factory. 16 porcelain fragments were recovered from this site, all of which are of blue- and-white plates and bowls from the *Wanli* period (1572-1620).

2-3. Real Monasterio de Santa Clara

The 'Claritas' was one of the first catholic orders to be established in Seville. The monastery was constructed in the 15th century above the old Muslim palace. The order was supported by donations and by the dowries of the nuns. Its economic power was considerable and some members of the orders acquired wealth by lending or investing in the trade with the New World, which was the biggest source of wealth in Seville during the colonial period. From this monastery many ceramics dating from the early modern to the modern period have

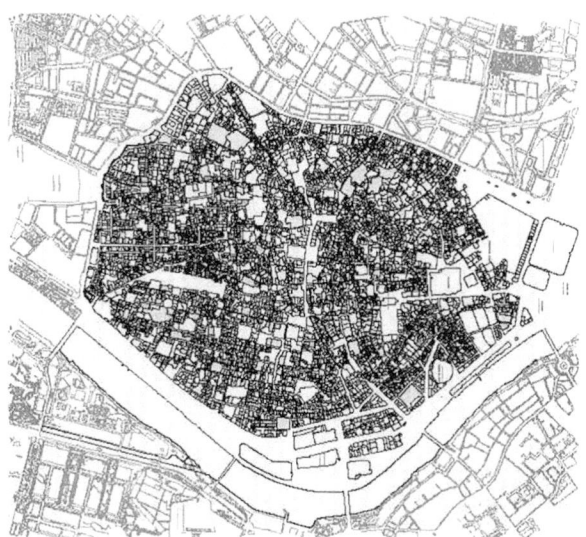

Map 3 Actual map of historical part of Seville
© Junta de Andalucía.

been excavated, and a few Chinese ceramics have been confirmed among the archaeological finds.

2-4. Altamira

The *Palacio de Altamira* is located in the area of San Bartolomé, facing the calle Santa Manría Blanca. After 1391 the property passed to Diego López de Zúñiga and the palace was constructed on his orders. From the 16th century the palace became the residence of Condado de Altamira and the building was renovated and modified according to the tastes of various owners until the 19th century. Many ceramics from several periods were excavated and one ceramic item of Chinese origin was found on the site.

2-5. La Florida

This site is located between La Puerta de Carmona and La Carne: between Mendez Pelayo Street, Luiz Montoto and La Florida. The site lies outside the city wall and the pavements, floor structure and well were excavated. These structures were possibly part of a residence belonging to the 15th century and may have functioned as an urban settlement. This settlement was constructed above the Medieval midden of the 13th and 14th centuries. A Roman necropolis and a Jewish cemetery were also excavated near this site. Ceramics from Roman to early modern times have been excavated and few Chinese ceramics were found.

2-6. Cuartel del Carmen

The earliest structure on this site is a Carmelite convent which dates back to the mid 14th century (1358) and continued for five centuries, extending its built space over the course of time. In the early 19th century the Carmelite order was expelled following the French invasion and the convent was taken over by the French military. The Carmelites returned later and the building was again used as a convent for a short period from 1812 to 1835. Subsequently it has been reused by the military until 1978. The building has been occupied and used continuously in its history and a good number of ceramics from Seville and other later ceramics, such as ceramic productions from *Cartuja* from the 19th century (1841). These are generally referred to as *loza industrial* (industrial ceramics) and they were found among other the artefacts. Most of the ceramic finds from this excavation were for daily use – cooking, eating and drinking.

The above are the major sites within the city where Chinese porcelain has been excavated and a total of 38 fragments have been found. The majority of these are of blue-and-white wares from the third quarter of 16th to the early 17th centuries; very few pieces are from the later periods.

2-7. Date phases

The mid 16th century to 1575

Blue-and-white bottle with scroll design (Photograph 27)

This is a small bottle with blurred scroll design painted on the body. The applied cobalt is very dark and tinted and the outline of the design is widely drawn over a buff-grey body, which is a common characteristic of the Jiajing period (1521-1567). However the roughness of this design may indicate a later period, possibly early Wanli. All in all it is most probable that this piece belongs to the third quarter of the 16th century and is probably one of the earliest types found in Seville.

Blue-and-white bowl with bird motif (Photograph 28)

A bowl with a pair of birds drawn on the exterior. An outline of a motif is drawn with slightly tinted cobalt. The wall is straight, a common characteristic of some bowls dated to the latter half of the 16th century.

PHOTOGRAPH 27: JINGDEZHEN BLUE-AND-WHITE BOTTLE EXCAVATED FROM THE REAL MONASTERIO DE SANTA CLARA, SEVILLE, DATED C. 1570-1575 (BELOW LEFT) (© MUSEO ARQUEOLÓGICO DE SEVILLA).

Chapter IV - Distribution of Chinese Ceramics and Asian Products in Spanish Society

Photograph 28: Jingdezhen blue-and-white bowl with bird design dated to c. 1550-1575, excavated from Cuartel del Carmen, Seville (© Museo Arqueológico de Sevilla).

Photograph 29: Jingdezhen blue-and-white sherds of plates excavated from Cuartel del Carmen (© Museo Arqueológico de Sevilla).

1575 to the late 16th century

Moulded blue-and-white plate with aquatic grass or landscape design (Photograph 29)

This is a thinly formed plate with lotus leaf or landscape drawn on the interior rim. The cobalt colour is much lighter than the previous type, which can be generally observed among pieces dated to the late 16th century.

Blue-and-white plate with phoenix design on the medallion (Photograph 30)

A plate with phoenix design on the medallion encircled by a wave design. The design is painted in bright cobalt. This is a type frequently found among the excavated pieces from the Zócalo area of Mexico City and also from the finds in Drake's Bay.

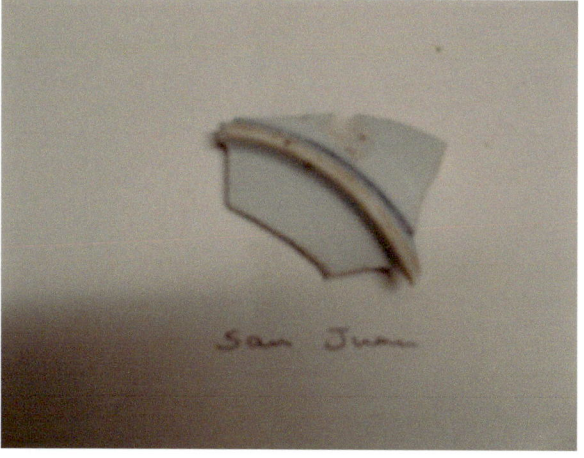

Photograph 30: Jingdezhen blue-and-white plate dated from 1575 to the late 16th century excavated from San Juan (© Museo Arqueológico de Sevilla).

PHOTOGRAPH 31: JINGDEZHEN BLUE-AND-WHITE SHERDS OF PLATE WITH TAIHU ROCKERY DESIGN EXCAVATED FROM CALLE SAN FERNANDO (© MUSEO ARQUEOLÓGICO DE SEVILLA)

The early 17th century

Blue-and-white bowl with deer and taihu rockery design (Photograph 31)

This bowl with a deer design on the exterior is a type common during the early 17th century. A stylized deer and wheel-like rock motif, which is generally called *taihu* rockery, is drawn using bright cobalt. This is an identical find to the one from the *San Diego* wreck which sank off Fortune Island in the Philippines following a Dutch attack in 1600, and also from the *Nossa Señora dos Mártires*, which sunk near Cascais, Portugal, in 1608. Thus this bowl can be dated to the early 17th century.

Blue-and-white bottle with long neck (Photograph 32)

A bottle with horizontal dot motif drawn on the narrow neck with tinted cobalt. A rough dot is painted between narrow segments. This is a type of *kraak* style applied on bottles with long and narrow neck and a globular body. The same type is commonly found on the *San Diego* wreck and also from the *Witte Leeuw* cargo, the Dutch ship which went down off St. Helena in 1613. (The ship was returning from Banten with cloves, nutmeg and pepper – and a hoard of 1,317 diamonds.) The diameter of the bottle neck is slightly larger than that found on the bottles of the *San Diego* and proportionally similar to those of the *Witte Leeuw*. Therefore it is probably dated between 1600 and 1630.

The late 17th to the early 18th century

Blue-and-white cup with scroll design from San Juan de Acre (Photograph 33)

A tall cup with scroll design on the body and geometric linear motif near the mouth rim. The mouth rim is unglazed, as if the glaze were wiped off with a cloth. This is a type produced during the Kangxi period (1661-1722) and generally the latter half of the 17th to the early 18th century. It was used as a chocolate or coffee drinking cup, or as a religious utensil in Mexico, and examples can be found from various sites, such as the Archbishop´s residence in Zócalo, Mexico City.

18th-century pieces

A plate with peony motif (Photograph 34)

A plate with large peony motif in the centre drawn with tinted cobalt over greyish body. This plate is formed from buff-grey clay, which is not a general characteristic of Jingdezhen production. Porcelain items from Jingdezhen are produced from clays free from impurities and the body is therefore normally pure white. On the contrary, ceramic production sites located in Fujian used buff-white to grey clays and these contain high levels of impurities. This plate is probably of Fujian or Guangdong production and not Jingdezhen. The peony flower-basket motif is very common on 18th-century pieces, and features

PHOTOGRAPH 32: JINGDEZHEN BLUE-AND-WHITE SHERD OF BOTTLE EXCAVATED FROM SAN JUAN DE ACRE, SEVILLE (© MUSEO ARQUEOLÓGICO DE SEVILLA)

Chapter IV - Distribution of Chinese Ceramics and Asian Products in Spanish Society

Photograph 33: Jingdezhen blue-and-white sherd of tall cup excavated from San Juan de Acre, Seville (© Museo Arqueológico de Sevilla).

Photograph 34: Possibly Guangdong (?) blue-and-white sherd of plate with flower-basket motif excavated from San Juan de Acre, Seville (© Museo Arqueológico de Sevilla)

on Chinese Imari porcelain and Guangcai (広彩) polychrome pieces. This blue-and-white plate is thus considered as a prototype of the later over-glazed, polychrome ceramics which were largely exported to many European countries. It is dated to the early half of the 18th century.

Tall blue-and-white Dehua cup from San Juan de Acre (Photograph 35)

A tall cup with floral design drawn with tinted cobalt over a buff-white body. This is identified as Dehua kiln (Fujian province) production from its buff, very soft clay and moulded form and tinted cobalt. Dehua kiln ware is traditionally known for its white ware production, especially the Guanyin figurines, and it was largely exported to Europe from the 17th and 18th centuries. Blue-and-white utensil wares were exported from the 18th century, although the quantities and quality cannot be compared to that of Jingdezhen. Cups were produced and exported to New Spain and Europe to meet the demands created by the new craze of chocolate and coffee drinking, and thus can be found at various excavation sites. This type from the San Juan de Acre site is dated to the 18th century.

Polychrome bowl (Photograph 36)

A European scroll motif is painted with red pigment over a mat-white body. Traces of gold pigment can be seen inside the scroll motif. Polychrome porcelain with such designs began to be produced apart from the 18th century and were exported in large quantities to all European markets. This fragment is a body part from a covered bowl. No similar pieces are found in other excavations, although from the design and shape it is identified as an 18th-century piece.

By studying the pieces excavated from the city of Seville it can be seen that the fragments date from the latter half of the 16th to the early 17th century; in particular the finds from 1550 to 1620 are most

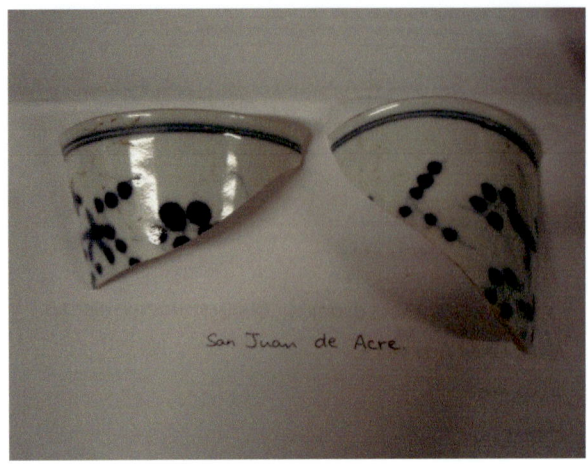
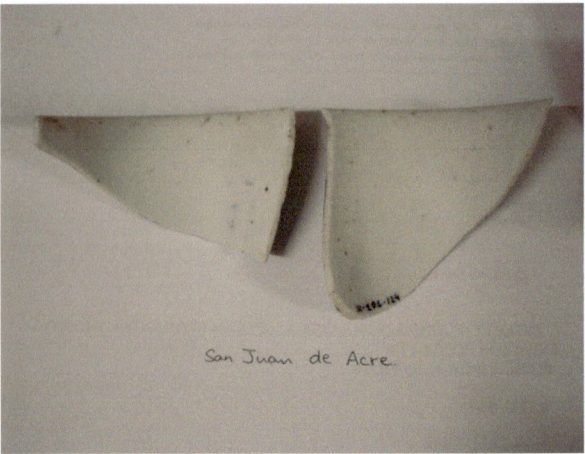

Photograph 35: Dehua blue-and-white cup excavated from San Juan de Acre, Seville (© Museo Arqueológico de Sevilla).

Photograph 36: Jingdezhen polychrome bowl with cover found at the La Florida site, Seville (© Museo Arqueológico de Sevilla).

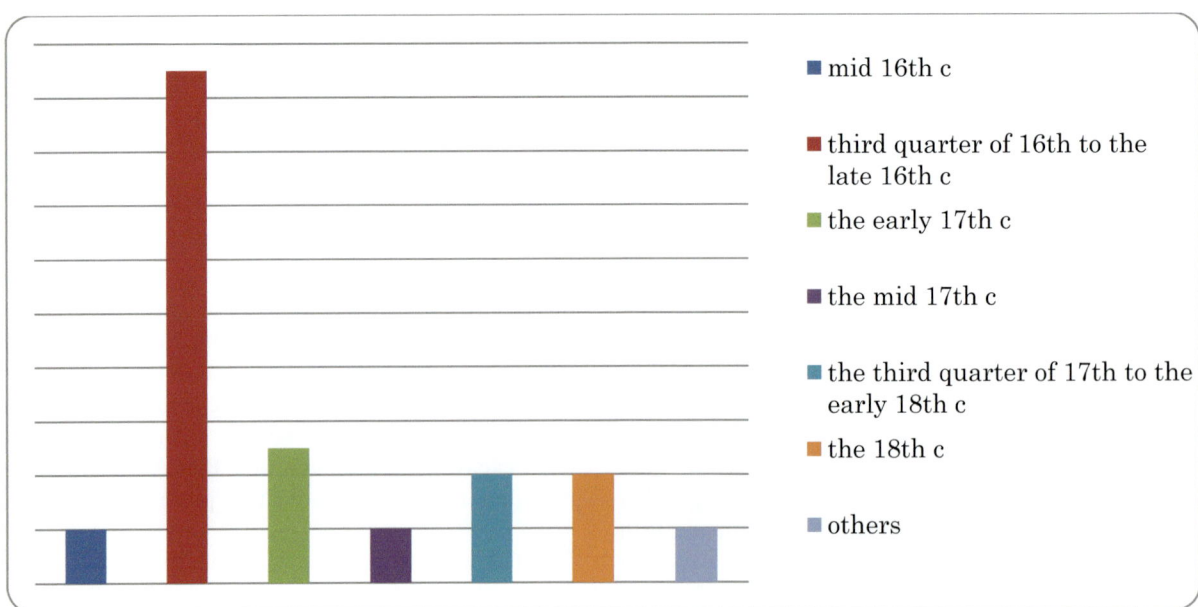

Figure 6. Quantity of excavated Chinese porcelain in Seville

varied and numerous compared to the other periods. Six types fall into this period, whereas four types were classified as being late 17th to the end of the 18th century – approximately 130 years. This indicates that the scale of the Asian trade in these products between New Spain and the European markets had changed over the course of time. The flow of Asian goods was more abundant during the latter half of the 16th to the early 17th century, which probably coincides with the scale of the Manila galleon trade.

2-8. Some short remarks

It is clear that from the third quarter of the 16th to the early 17th century the quantities of Chinese porcelain exports to Spain from Veracruz increased. This is a common tendency compared to other European markets, such as Portugal, and it also coincides with the situation in Mexico, where the ceramic trade with Asia was at its most dynamic during this period.

Considering the actual quantities of Chinese ceramics brought to Seville, it is clear that the amounts were extremely small compared to those of other European countries, such as Portugal and Holland, where a single ship could carry a much larger number of ceramics for trade purposes. This is clear from the cargo register in the *Archivo de Indias* from 1589 to 1601 (see the previous chapter), and also from the ceramic materials of excavation sites in Seville. The total number of 38 fragments from all the sites within the old city is too small to be considered as material distributed in the market. Moreover, this theory is supported by the archive, which mentions that some of the ceramic cargoes were personal presents or belongings. The archaeological fact that almost all of the fragments excavated were from bowls and plates, as well as the historical document that solely refers to plates and *escudillas* (bowls), match perfectly and thus indicate that the imported Chinese ceramics were for table wares. The nature of these ceramics in Spanish material culture was for use as table wares, which were only distributed among a limited social class who appreciated Chinese porcelain as luxury items. This hypothesis can be substantiated by the possibility that elements of Spanish society may have appreciated silverware more than Chinese ceramics because of the abundant imports of silver from America. According to the study of Jesus Aguado de los Reyes, the tangible assets of the population of Seville in the 17th century included coin, silverware, and jewellery as luxury items, but ceramics did not appear to feature in any list of property for any social class.

It can be said that the demand for, and distribution of Chinese ceramics among Spanish society was less affected by the Portuguese markets, where a large demand for Chinese ceramics began earlier, as a result of its direct trade with Asia, using Macao as a commercial base. In effect the majority of Chinese ceramics brought to Spain came from New Spain, using its own trade route, and was hardly ever linked with other material cultures. The large demand for Chinese ceramics in the Netherlands was stimulated by Portuguese trade, as witnessed by the Dutch capture of the *Sanyago* and *Santa Catalina*, ceramic cargoes of which were sold in auction in Amsterdam and the volume of trade in ceramics by the Dutch greatly increased subsequently. This was never the case in Spain and it must be concluded that the influence of Asian art on Spanish culture was very limited. On the other hand, it is very curious that Chinese ceramics had some influence on *Sevillana* ceramics of blue-and-white wares and several examples were found among the excavated materials. A sherd in Photograph 37 shows a close similarity with Chinese porcelain found in the *Vung Thau* wreck, sunk in the 1670s.

PHOTOGRAPH 37: SEVILLE BLUE-AND-WHITE WARE WITH CHINESE INFLUENCE (LEFT) (© MUSEO ARQUEOLÓGICO DE SEVILLA). JINGDEZHEN BLUE AND WHITE PLATE FROM THE *VUNG THAU* (RIGHT) (© MUSEO NACIONAL DEL ARTE DECORATIVO).

3. Chinese Porcelain in Lisbon and the Galician Coast

Although in the previous context the present author has referred to Chinese porcelain excavated in Spanish territory, it seems appropriate to separate the Galician province in northwest Spain as it is adjacent to Portugal and connected along the coast; and from the earlier period it maintained culturally, linguistically and commercially a very close relationship to Portugal. Coastal Galicia was considered peripheral, and transportation from central Spain was hindered by the mountains. Therefore it is perhaps more logical to study coastal Galicia in the overall context of the Atlantic coastal trade with Portugal and the Netherlands. On the other hand, a large number of the pieces found in Galicia is concentrated in Santiago de Compostela, a famous pilgrimage city where wealth and culture were concentrated. In contrast with the coastal port-cities of Galicia, Santiago de Compostela was connected by the pilgrimage route in all directions since medieval times. This land route towards the interior makes Santiago de Compostela a Galician city of a very different nature (see Map 4).

3-1 Excavated ceramics from Bayona

13 pieces in total were excavated from a site in the city of Bayona, which is located north of Oporto (Portugal) along the Atlantic coast, a little south of Vigo (see Map 3). The excavations were carried out in 1992 adjacent to the fort know as Monterreal, which protected the old town of Bayona. This old centre, still called Vila Vella, is first recorded in a document of 1201 as 'Baiona'. The city included a monastery, church, and several habitation; it remained an urban site until 1665. As a result of its geography, the city was always involved in territorial disputes between Portugal and Galicia. Moreover, the city, facing the Atlantic, had the great advantage of being accessible to many English, Irish and Dutch merchants during the reign of Philip II. The port shared a monopoly with the port of Pontevedra with regard to the unloading of imported goods within Galicia. In 1663 war broke out against Portugal and in 1665 the governor and the captain general of Galicia, Luis Poderico, ordered the destruction of 136 houses in Vila Vella to defend Fort Monterreal from Portuguese troops. Therefore the dating

Map 4: Map of Spain and Portugal.

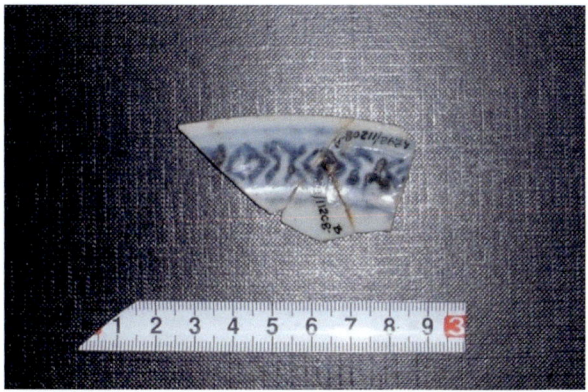
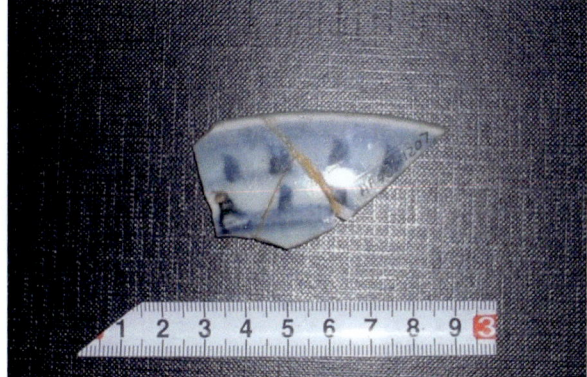

PHOTOGRAPH 38: JINGDEZHEN BLUE-AND-WHITE SHERD OF PLATE EXCAVATED FROM MONTERREAL.©MUSEO DO MAR

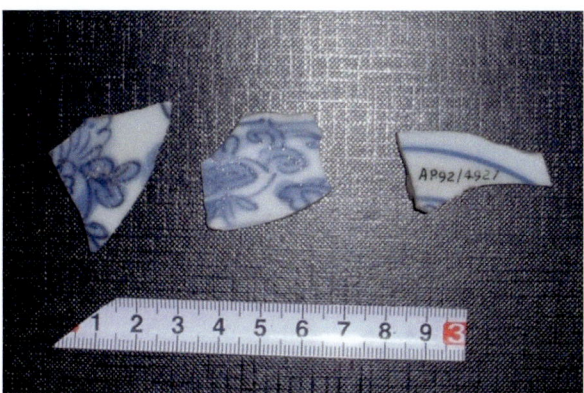
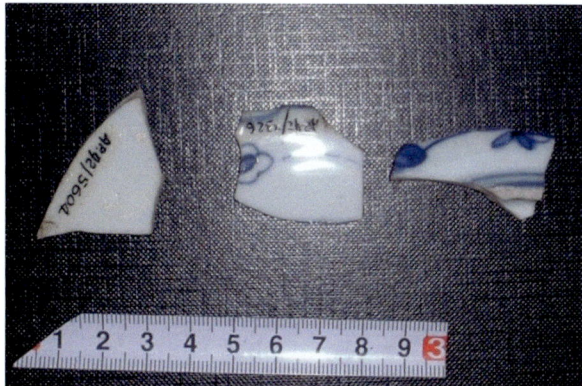

PHOTOGRAPH 39: JINGDEZHEN BLUE-AND-WHITE PLATE WITH *LINZHI* MOTIF ON THE RIM EXCAVATED FROM MONTERREAL.[162]

of the last cultural layer in this excavation is considered to be around 1665.[161]

Blue-and-white plate with geometric design (Photograph 38)

A band of geometric design drawn with rather dark cobalt on the flat rim; a blurred dotted design on the exterior of a lion, or qilin, is frequently painted in the centre. The same type can be seen among the excavated pieces from Alfama, Lisbon. It is thickly potted compared to other *kraak*-type wares, and the base colour is dark blue to grey, although it is produced in Jingdezhen. This type can be dated from the early to the third quarter of the 16th century, which suggests it is probably the earliest piece among the pieces found in Galicia. The plate normally shows a lion chasing a ball [ball or bull??] drawn on the central medallion.

Blue and white plate with *lingzhi* design (Photograph 39)

A thinly potted plate with a stylized *lingzhi* (?) design drawn on the rim. This *lingzhi* (靈芝) design (or sacred fungus symbol) is frequently seen on the exterior design of many *kraak*-type plates and bowls, although most are drawn in a very rough and stylized way as a wavy design on the exterior, as can be seen on the plates and bowls from the *Witte Leeuw* wreck of 1613.[163] It is rare, therefore, to find the *lingzhi* symbol on the interior, especially with this rather neat brushwork. A similar design can be seen on the exterior of one of the plates dated to 1573-1619 in the Colecçao Aamaral Cabral in Portugal,[164] and also on a fragment with a similar motif from Vigo (H95/8476). Based on its exterior floral design and comparison with pieces from the *Witte Leeuw* cargoes, the find can be dated to the late 16th century.

Blue-and-white bowl with lotus leaf and egret design (Photograph 40)

A fairly tall bowl with a repeated floral spray design on the exterior and lotus leaf and egret design on the interior mouth rim. The lotus leaf and aquatic grass with egret design appears frequently from around the mid 16th to early 17th century, at the latest, and lasting about half a

---

[161] Vicente Caramés Moreira y Fátima Cobo Rodríguez, Porcelana chinesa da dinastía Ming Procedente do Parque da Palma de Baiona, *Castrelos*, nº. 13, Vigo, Museo Municipal de Vigo Quiñones de León, 2008, pp. 96-106.
[162] Ceramic finds originally belonged to Vicente Caramés Moreira, which now are stored in Museo do Mar.

[163] Pijl-Ketel, C. L., van der (ed.), *The Ceramic Cargoes of The Witte Leeuw*, Amsterdam, Rijksmuseum, 1982, p. 62.
[164] Pinto de Matos, Maria Antonia, 'A porcelana chinesa na Colecçao Amaral Cabral', *Azul e Branco da China*, Lisbon, Loja das Ideias, 1997, p. 142.

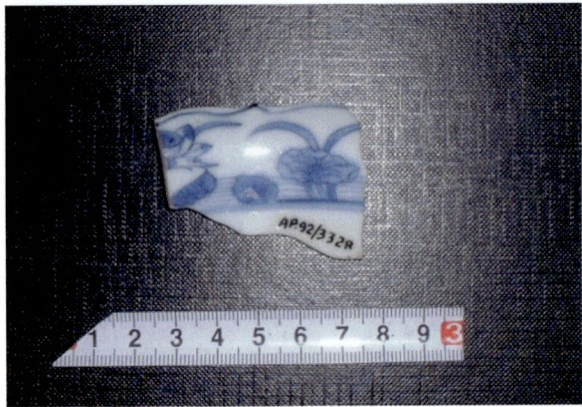
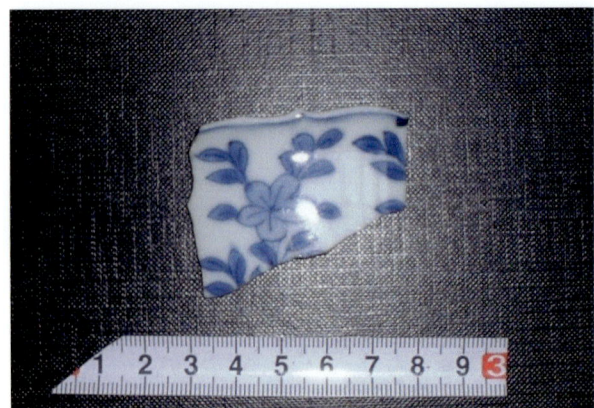

PHOTOGRAPH 40: JINGDEZHEN BLUE-AND-WHITE SHERD OF BOWL WITH EGRET AND FLORAL DESIGN EXCAVATED FROM MONTERREAL.

PHOTOGRAPH 41: JINGDEZHEN BLUE-AND-WHITE SHERD OF PLATE WITH DEER AND PINE MOTIF EXCAVATED FROM MONTERREAL.

From its quantity, it was probably massively produced and exported via Manila to New Spain and also via Macao to Lisbon. A similar piece has been found in Drake´s Bay, and is associated with San Augustin pieces; therefore, it can be dated to the late 16th century.

Blue-and-white plate with deer and floral design (Photograph 43)

A late *kraak*-type blue-and-white Jingdezhen plate with a deer design on the centre. It is thinly potted and the tone of cobalt is somewhat light, roughly expressing two

PHOTOGRAPH 42: JINGDEZHEN BLUE-AND-WHITE SHERD OF PLATE WITH EGRET MOTIF EXCAVATED FROM MONTERREAL.

century. However, neatly drawn motifs, such as the one on this piece, is more likely to be dated to the late 16th century. A similar blue-and-white bowl with a repeated deer design on the exterior can be found among the *San Diego* pieces (1600), although this floral spray design is possibly slightly earlier.

Blue-and-white plate with deer and pine design (Photograph 41)

A *kraak* blue-and-white plate with a deer and pine motif drawn on the central medallion. The deer, pine and *taihu* rockery design is a very common *kraak* motif on late 16th-century plates. This type is present among the Drake's Bay finds, indicating that this piece is dated from the late 1575 to the early 17th century.

Blue-and-white plate with egret and eight treasure symbols (Photograph 42)

This type is very common among the excavated pieces in Mexico City and also from the Portuguese collection.

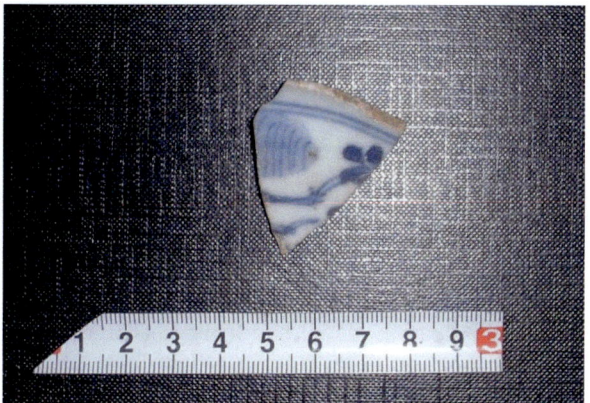

PHOTOGRAPH 43: JINGDEZHEN BLUE-AND-WHITE SHERD OF PLATE WITH DEER DESIGN EXCAVATED FROM MONTERREAL.

front legs of a deer with floral motif beneath. The centre medallion is encircled by double lines. This type is found among the *Hatcher* cargo (1643), and thus these pieces can be dated to the first half of the 17th century.

The majority of the pieces excavated from Bayona are small fragments; their dates are concentrated around the third quarter of the 16th century. This indicates that the pieces are from one refilled layer. Although the site continues until the mid 17th century (1665) no ceramics from the late 17th century were found.

3-2 Excavated ceramics from Vigo

There have been 13 Chinese ceramics excavated from the former military hospital and the 16th-century ex-convent of Vigo. The excavation was carried out in 1995 within the present city (see Map 4). The military hospital was constructed in the 19th century over a structure that was first used as a Franciscan convent in 1553. The convent was completely destroyed by a fire that broke out when Francis Drake attacked Vigo in 1589. Therefore, the latest dating of the ceramics found at this site can be to around the late 16th century.

The excavated Chinese ceramics found near the wall structure of this convent are associated with late 16th- to early 17th-century German salt-glazed stoneware. These grey- and blue-glazed stonewares produced in Westerwald (Germany) are found in the excavations of major cities in Galicia, indicating commercial relationships between this region and northern European cities.[165] The city of Vigo faces the Atlantic and has been an important port-city up to the present time. In 1571 a document records that 700 families lived there, and ships from Brazil, America and Cape Verde entered the port there in preference to Bayona for security reasons.[166] It is also known that the port was permanently in competition with Bayona and La Coruña, where many Dutch and English merchants unloaded their cargoes.[167] The importance of this city is proved by the attack of Francis Drake in the 16th century and also by the famous battle of Rande that took place offshore from this port between Spanish and French ships in 1702, leaving many wrecks still in the waters there today.[168]

Blue-and-white plate with floral design (Photograph 44)

A *kraak*-type plate with a floral design on the medallion, surrounded by a light blue band and scroll design. This scroll design appears from the mid 16th century until around 1600 at the latest. This motif is absent from the *San Diego* and *Witte Leeuw* finds, and it is probable therefore that this type was produced in the late 16th century.

Other pieces are mostly the same as those types found in Bayona excavation. They include a blue- and-white plate fragment with *lingzhi* symbol, a plate with an aquatic grass design on the flat rim, and a plate with eight treasure symbols on the rim and a phoenix design on the medallion. The components of the ceramics excavated from this military hospital site in Vigo are roughly similar to those from Bayona. A common characteristic of the ceramics from both sites is that a major part was produced in Jingdezhen, being *kraak*-type plates from the third quarter/late 16th century. Each of the fragments is small, indicating that these pieces are also from a refilled layer from the same period.

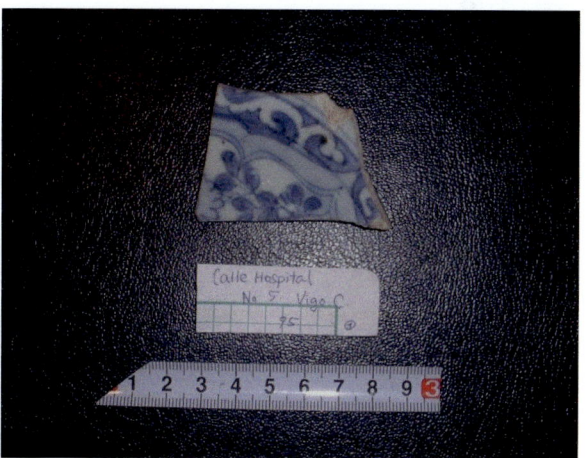

PHOTOGRAPH 44: JINGDEZHEN BLUE-AND-WHITE SHERD OF PLATE EXCAVATED FROM CALLE HOSPITAL, VIGO.[169]

---

[165] Manuel Xusto Rodríguez, *Hasta el Confín del Mundo: Diálogos entre Santiago y el Mar, Vigo*, Gráficas Varona, 2004, p. 240.
[166] Mª del Carmen Gonzales Muñoz, ‹Vigo y Su Comarca en los Siglos XVI y XVII›, *Vigo en Su Historia*, Artes Gráficas Galicia, Vigo, 1979, pp. 153-276.
[167] Mª del Carmen Gonzales Muñoz, *op. cit*, pp. 153-276.
[168] González Fernández, Juan Miguel, 'El escenario: Vigo y su ría en torno a 1702', *Rande 1702: arde o mar*, Museo do Mar de Galicia, Vigo, 2002, pp. 117-133. Rodríguez-Villasante Prieto, Juan Antonio, 'La Defensa de la Ría de Vigo, Campaña de 1702', *Rande 1702: arde o mar*, Museo do Mar de Galicia, Vigo, 2002, pp. 135-157.
[169] Ceramic founds from Calle Hospital, Vigo belongs to Angel Acuña

3-3. Excavated ceramics from Santiago de Compostela

This city is well known as a Catholic pilgrimage destination via the 'Camino de Santiago', an important 9th-century pilgrimage route (see Map 4). Today many pilgrims travel the 800km from France to reach the city´s cathedral, where the relics of Saint James are said to rest. The site is located about 70km east from Vigo, towards the interior. The excavation carried out during the last century proves that the city, especially around the cathedral, was already inhabited during the Roman period, and there exists a necropolis underneath the city cathedral. The previous study indicates that the population of this city can be calculated to 10,000, including the surrounding towns and villages in the mid 16th century. The city was one of the most populated in Galicia in this period, being a religious and academic provincial centre.[170]

Over 40 fragments of Chinese ceramics, mainly from the 16th and 17th centuries, were excavated from a structure now called the 'Casa del Deán'. This former residence first appears in documents in 1706 when Domingo de Arellano allowed the land of two houses in Rúa do Vilar, which runs from Plaza de Platería, to be used for a new house for a canon ('casa para vn señor canónigo').[171] Plaza de Platería is located just in front of the rear entrance to the cathedral, where many of the silverware shops were to be found. From the location of the site it is considered that the area has always been inhabited by, or belonged to the nobility of this religious city. Before the construction of Casa del Deán, which was completed in the 18th century, the land belonged to Mencía de Andrade, as documented in 1571. According to the documents the land was used as a private residence throughout the 16th century.[172] All the ceramics were found in a refilled layer underneath the construction of the 18th-century residence.

Blue-and-white bottle with lion design (Photograph 45)

A bottle with the motif of a lion chasing a ball with *ruyi* cloud floating above and a wave design drawn close to the foot-ring. The piece has lugs on both sides. The four Chinese characters 「永保長春」 (meaning eternal preservation of the long Spring) are written on the base. The stylized *ruyi* cloud is similar to that known from the Jiajing period, as well as the lion chasing a ball [ball or bull?] motif frequently found on Jiajing plates and bowls. The base mark is similar to that seen on a late Jiajing plate in the Amaral Cabral collection.[173] No excavated example of this type is known but from the style and mark it can be dated to the mid 16th century.

Blue-and-white plate with floral design (Photograph 46)

A *kraak*-type plate with panels decorated with twig and leaves. The equal segments are one of the characteristics of the early *kraak* forms and are seen among the excavated pieces from the Drake´s Bay finds and therefore it can be dated to somewhere in the third quarter of the 16th century.[174] A similar piece was excavated from the Templo Mayor in Mexico City.

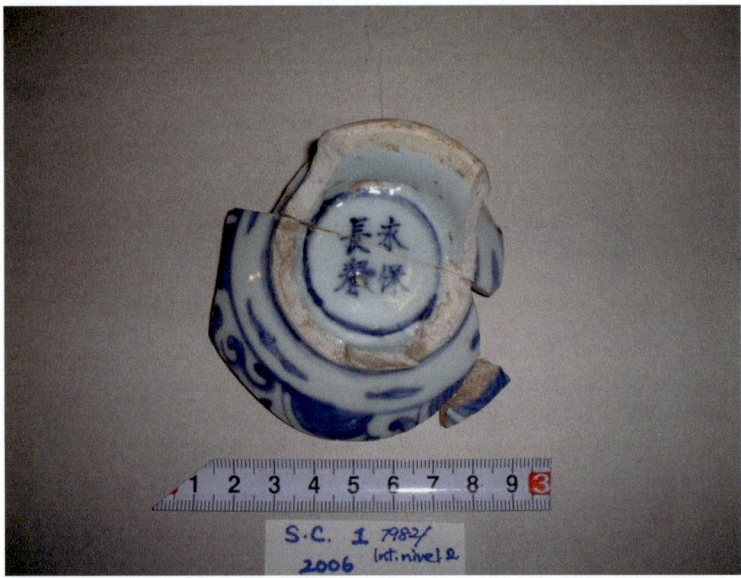

PHOTOGRAPH 45: JINGDEZHEN BLUE-AND-WHITE SHERDS OF BOTTLE DATED TO THE MID 16TH CENTURY EXCAVATED FROM CASA DEL DEÁN.[175]

---

[170] Maria del Carmen Gonzales Muñoz, Galicia en 1571: Poblacion y Economia, *Edicios do Castro Serie Liminar historia*, A Coruna,1982, pp.60-69.
[171] Miguel Taín Guzmán, *La Casa del Deán de Santiago de Compostela,La Coruña*: Diputación de la Coruña, 2004, pp.17-47.
[172] Miguel Taín Guzmán, *op. cit*, pp.17-47.
[173] Maria Antonia Pinto de Matos, *op. cit*, pp. 74-75.
[174] Clarence Shangraw and Edward Von der Porten, *The Drake and Ceremeño Expeditions´ Chinese Porcelains at Drake´s Bay, California, 1575 and 1597*, Santa Rosa Junior College and Drake Navigators Guild, 1981, p. 25.
[175] Ceramic finds from Casa del Deán belonged to José Suarez Otero

CHAPTER IV - DISTRIBUTION OF CHINESE CERAMICS AND ASIAN PRODUCTS IN SPANISH SOCIETY

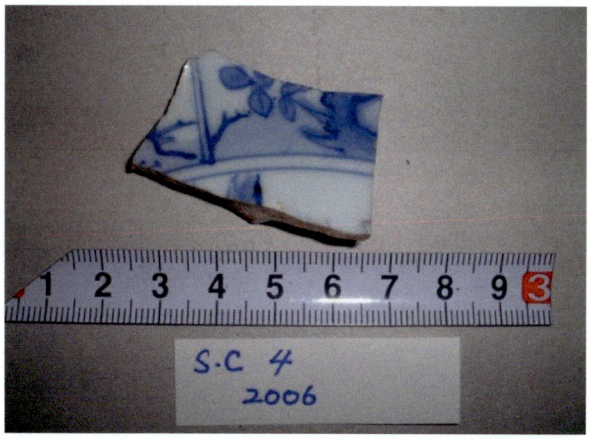
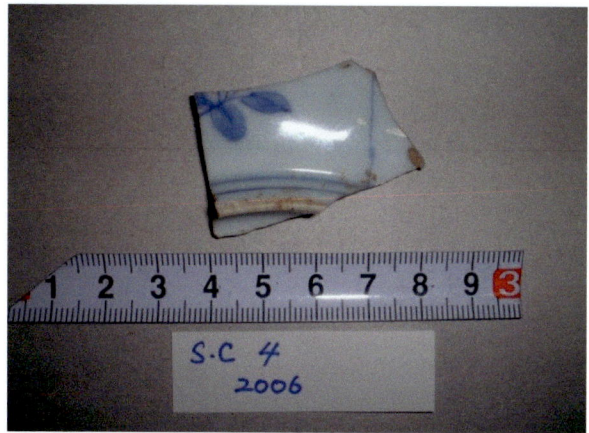

PHOTOGRAPH 46: JINGDEZHEN BLUE-AND-WHITE SHERD OF PLATE EXCAVATED FROM CASA DEL DEÁN.

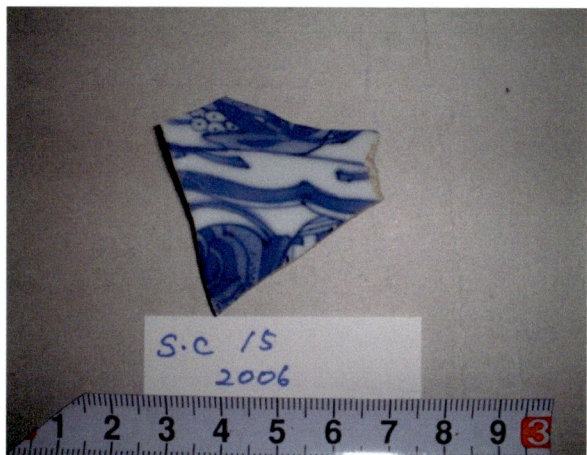
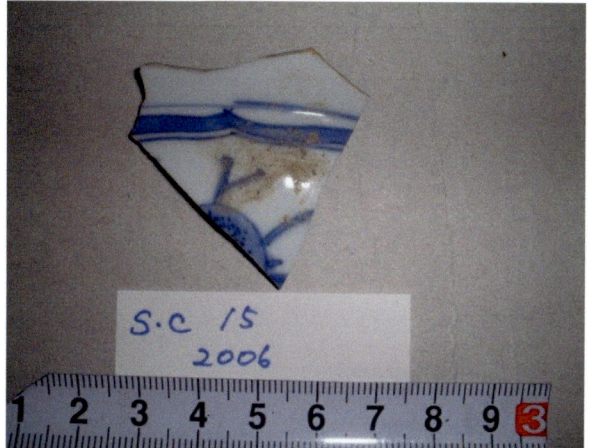

PHOTOGRAPH 47: JINGDEZHEN BLUE-AND-WHITE SHERD OF DEEP BOWL (*KLAPMUTSEN*) EXCAVATED FROM CASA DEL DEÁN.

Blue-and-white *klapmutsen* with Taoist symbol design (Photograph 47)

Klapmutsen is a term applied to a form with rounded walls and flattened rim; it is a shape in between a traditional Chinese bowl and a *kraak* plate. The motif drawn is part of a Taoist symbol and can be seen among the finds from the *Witte Leeuw*. This fragment can be dated c. 1610 to 1650 at the latest.

Blue-and-white moulded plate (Photograph 48)

A *kraak*-type moulded plate with the segment decorated with dots and diaper patterns. The larger oval panel is partly decorated in a floral design, although from this small fragment it is difficult to identify the whole design. The underside is decorated with a roughly drawn stylized *lingzhi* and oval panel. This typical *kraak* style with panels on the cavetto is not as heavy as the *Witte Leeuw* pieces, although the date falls into the first decade of 17th century.

Blue-and-white plate (Photograph 49)

This fragment only allows us to see the dot design which normally divides the panel on the cavetto of *kraak* types.

The brushwork is rough, as the division panels are only expressed by a series of dots without the tassel. This design started to be produced in the late 16th century and is present in the finds from *San Diego*, although the dots are drawn more carefully with tassels below them. This fragment, judging by its style, is closer to some of the *Hatcher* finds, or even later.[176]

It is clear that the ceramics excavated from Casa del Deán are more abundant and varied in form than the finds from Bayona and Vigo. The ceramics included white wares, mostly plates, although the fragments were not large enough to determine dates. The large quantities of finds from this religious city, where elites and clergies resided, suggest that these Chinese ceramics were closely connected to the wealth of the residents. In any event, the items excavated from this site include pieces from around the mid 16th century to the 1630s. Some of the types coincide with those of Bayona and Vigo, although some of the 17th-century pieces are similar to finds from the *Witte Leeuw*.

---

[176] Pijl-Ketel, C. L., van der (ed.), *op. cit*, p. 82.

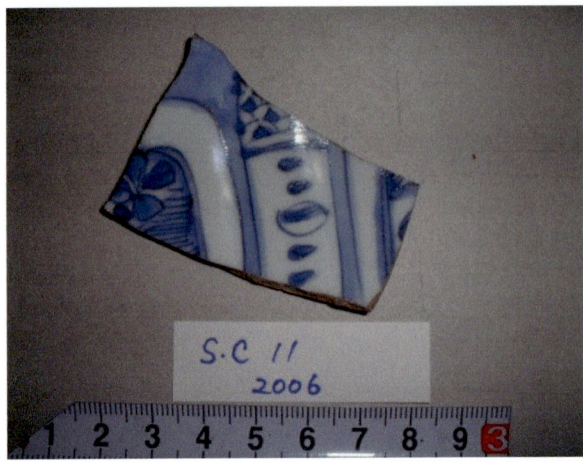 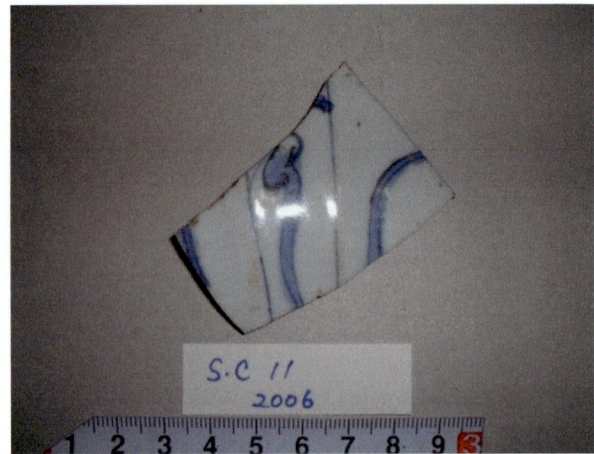

PHOTOGRAPH 48: LARGE JINGDEZHEN BLUE-AND-WHITE *KRAAK*-TYPE MOULDED PLATE EXCAVATED FROM CASA DEL DEÁN.

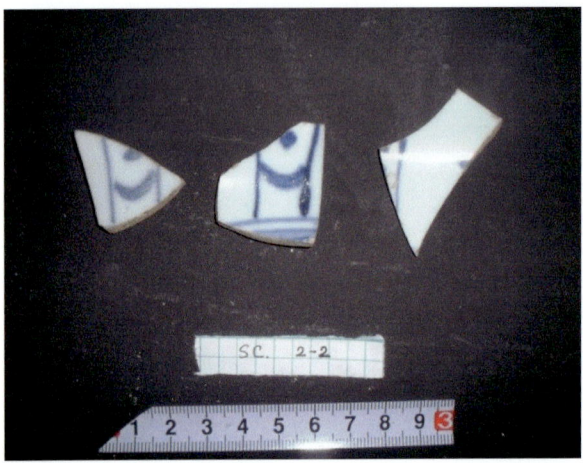 

PHOTOGRAPH 49: JINGDEZHEN BLUE-AND-WHITE SHERD OF PLATE EXCAVATED FROM CASA DEL DEÁN.

3-4. Excavated ceramics from Orense

Orense is an interior city 100km east of Vigo, on both banks of the Río Miño. The origin of this town dates to the Roman period and the site was always of strategic importance in terms of defending against the exterior threats that came up from the Rio Miño. During the 17th and 18th centuries many religious buildings were constructed by several orders, such as the Jesuits, within the city (see Map 4). Chinese ceramics were excavated from the fort site of Maceda (Castillo de Maceda), which is located 30km from the city of Orense. The character of the fort is more residential palace than mere defensive fort.

Blue-and-white plates (Photograph 50)

Three small fragments of blue-and-white plates were excavated from this site, all from Jingdezhen provenance. These fragments are from *kraak*-type plates, although it is difficult to ascertain the original designs. The fragment on the right in the photograph is from a plate which has a division with a dot design on its cavetto and an animal motif on the central medallion. The style is similar to the *kraak* type of the early 17th century (*San Diego* and *Witte Leeuw* finds). Although the dating of the other two pieces is more difficult to define, they are both cavetto parts with divisions, and can be dated approximately to the late 16th/early 17th century.

PHOTOGRAPH 50: JINGDEZHEN BLUE-AND-WHITE SHERDS OF PLATES EXCAVATED FROM THE MACEDA SITE.

CHAPTER IV - DISTRIBUTION OF CHINESE CERAMICS AND ASIAN PRODUCTS IN SPANISH SOCIETY

3-5. Significance of Chinese ceramics in the Atlantic coastal trade

Chinese ceramics found in Galicia are not abundant in number and we lack therefore sufficient information to analyse fully the pieces and place them into historical context. From what we have seen from the excavated pieces, however, it is possible to say that the dates and types from the interior areas vary from those from the coastal sites. The examples from Bayona and Vigo are similar in type and date and most of the pieces were similar plates from the second half of the 16th century, whereas some of the finds from Santiago de Compostela, although from the same period, at the same time include later items – such as *kraak* wares. On the other hand, the pieces excavated from Orense are all from the later period, probably from the early 17th century. The artefacts from Santiago de Compostela (Casa del Deán), especially, are more varied and the dates cover a wider range. The datings range from the mid 16th to the first half of the 17th century. One of the possible interpretations of this chronology is that the two port-cities were strongly connected with the Atlantic coastal trade and the pieces were imported via Portugal. On the other hand, Santiago de Compostela, being one of the cities where wealth was concentrated in Galicia, could have been the recipient of ceramics distributed from other centres in Europe, such as Amsterdam, although the closest port would still be on the Atlantic littoral. The same theory can be applied for the city of Orense, where many religious orders had their churches, monasteries and convents, as well as the fort of Maceda.

The Atlantic coastal trade between Galicia and Portugal has existed since the medieval period. Galicia exported fish (mainly sardines) and ship construction materials, and imported salt from Portugal. The ports of La Guardia, Bayona and La Coruña were all participating in this trade. By the early 16th century the Galician merchants were also active in Portuguese territories, establishing their communities in Lisbon. The largest trade ports in Portugal were Oporto, Viana do Castelo and Aveiro (map 4). From the 1550s to the 1560s, some Galician communities in Viana do Castelo were engaged in fishing and other trades, and many others were engaged in trade with Asturias and Viscaya (the Basque region) as middlemen.[177] From the mid 16th century Galicia imported sugar from Portugal, which was produced in the Portuguese colonial plantations. Sugar was distributed within the region or passed through on its way to Asturias and the Basque region, where there was a larger market.[178] Moreover, many ships from Madeira and Brazil, loaded with sugar, headed directly to Galicia clandestinely. An interesting fact is that in the first few decades of the 17th century, when the Inquisition began in Portugal, in order to escape '*caçada no Burgo aos homens do dinheiro*' ('hunt for men of wealth') new Christians or converted Christians in Oporto, who had probably made fortunes from the sugar trade, had their lands and other properties confiscated. Some of these Jewish merchants, who usually had contacts in Galicia, took refuge in Galicia and Asturias.[179]

4. Classification of ceramics from Lisbon

Chinese ceramic materials have been excavated from various sites in Lisbon to date and a certain quantity of Chinese ceramics has been found together with the Lisboetan ceramics, Italian Majolicas and Spanish ceramics. The Chinese ceramics classified and discussed in this study are from the excavation site in Alfama, the old heart of the city of Lisbon (see Map 5). It is a structure of a medieval wall built by King Fernando in 1375 and also a segment of a reinforcing structure that formed the lower part of the fountain called, 'Chafariz dos Cavalos', and is dated between the late 16th and early 17th century. The site also includes the exterior part of the above-mentioned wall which is adjacent to the river front and an area of debris. The ceramic materials excavated here seem to be related both to the wall element and the fountain structure, although most of them fall within the first half of the 16th century. Recently the CHAM national research institute (Centro Histórico e de Alejam Mar) has been carrying out several underwater archaeological explorations and many items of Chinese porcelain have been found from the wreck of *Nossa Senhora dos Mártires* which sank off Cascais, near Lisbon, in 1608.[180]

4-1. The earliest type of ceramics

A) A large and thickly thrown plate with geometric design on the flat rim and a lion or qilin motif on the central medallion. The exterior wall has two variations: one has repeated circles of a tide motif near the rim, with the wall normally left without design, and is carved in horizontal lines; the other has a variant of a half-moon motif repeatedly drawn. Some of them have a foliated and others a straight rim. The bottom is carved out roughly with the 'chatter' marks left. It is not clear if these differences are reflections of different dating, although most probably they all fall into the first decades of the 16th century. They are also found in the Philippines and other Asian sites. Fragments of this type are abundant at this site in Alfama compared to pieces of other types and thus may be considered as a popular form in Portuguese markets during the early 16th century (Photograph 51).

---

[177] Aurélio Oliveira, Do Porto a Pontevedra (XIII-XVII): Os Tratos Maritimos no Noroeste Peninsular. Do Porto à Galiza e a Pontevedra, *Actas do Simposio de Historia Marítima do Século XII ao XVI*, Calo Lourido, Francisco (cood.), Pontevedra, Graf. Duher, 2003, pp. 111-138.
[178] Aurélio Oliveira, *op. cit*, pp. 111-138.
[179] Aurélio Oliveira, *op. cit*, pp. 109-131.
[180] Pavilhao de Portugal, Expo'98, Lisboa, 'Nossa Senhora dos Martires: The Last Voyage', Lisbon, 1998.

Portuguese Intervention in the Manila Galleon Trade

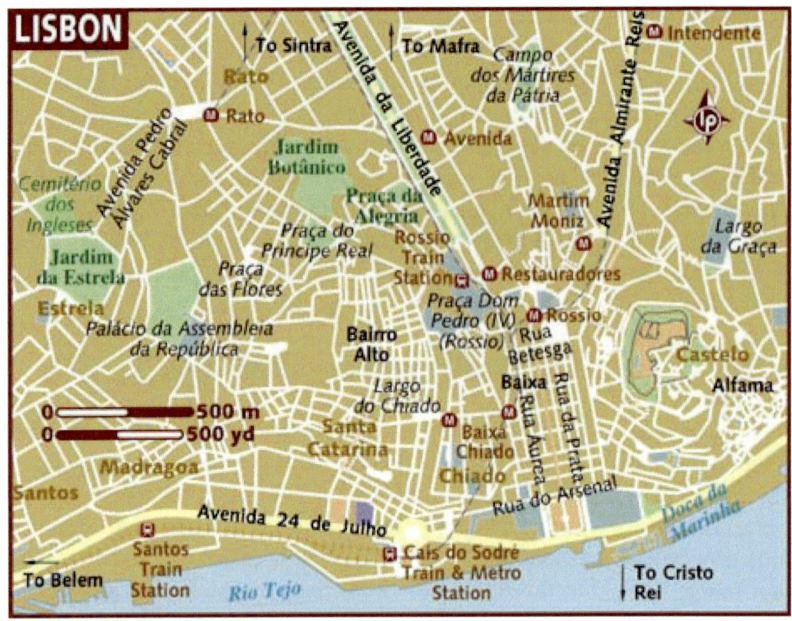

MAP 5: LISBON.

The frequency of these bowls found in Asia indicates that they were produced extensively to meet Asian demand but were also exported to Portugal.

These bowls are not often found at this excavation site in Lisbon and therefore were probably types no longer popular locally. The design and size of these bowls were clearly aimed for Asian tastes and are not abundantly found among heirloom pieces in Portugal. They may be the earliest types that were traded by the Portuguese (Photo 53).

4-2. The first half of the 16th century

This piece may have been produced in the provincial Jingdezhen area, as indicated by its rough clay and tinted cobalt. As mentioned previously, the same type is present among the pieces found in Bayona.

B) Another type which can also be dated to the early 16th century is a bowl with a conch shell drawn on the interior centre with a thick line. This type is also frequently found in many excavations in Asia and is normally dated to the latter half of the 15th century. It is likely that, together with the previous type, this style of bowl was produced until the early 16th century (Photograph 52).

C) A bowl with floral spray design on the exterior. This is a type which falls into the same category as the previous one and dates to the latter half of the 15th century in Asia.

E) A large plate with floral design on the interior and a cross mark on the base. This is a type which has been found in excavations at Penny's Bay, Hong Kong. Other pieces were also found on Shangchuan Island. This interesting cross mark has been suggested originally as the emblem of the Knights Templar, which was later adopted by the Portuguese Order of Christ. The cross mark was also used on Portuguese coins during the late 14th to early 15th century.[181] This form was most probably produced for the Portuguese market and can be dated to the first decades of the 16th century according to the excavation report.[182] Chinese ceramics exported to Portugal during this period were mostly consumed within its territory, since very few examples are found in other European countries (Photograph 54). This piece shows an interesting link between Hong Kong, Shangchuan, and Lisbon, and seems to prove that there existed a

 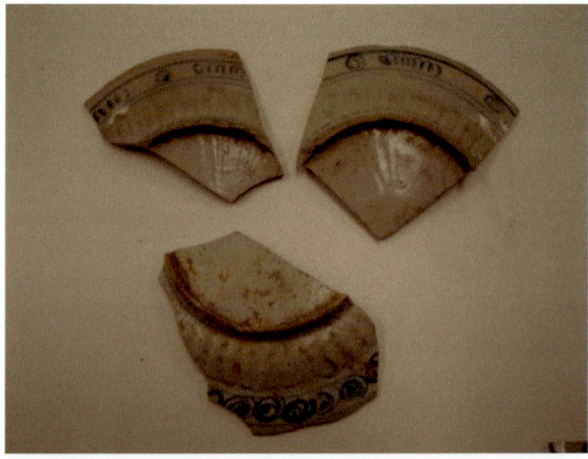

PHOTOGRAPH 51: CHINESE CERAMIC SHERDS EXCAVATED FROM ALFAMA (© MUSEU DA CIDADE).

---

[181] Wei Huang and Qing Huang, Guangdong Taishan Shanchuandao Huawanpin Yiji Chutu Jijijixiangguan Wenti, Wenwu, Diwuji, Beijing, 2007, pp. 78-88.
[182] Wei Huang and Qing Huang, *op. cit*, pp. 78-88.

Chapter IV - Distribution of Chinese Ceramics and Asian Products in Spanish Society

Photograph 52: Chinese ceramic sherd with conch shell motif in the centre excavated from Alfama (below right) (© Museu da Cidade).

private, or 'contraband', trade with the Portuguese and Chinese along Guangdong coastal area during the early 16th century.

F) A large bowl with phoenix or peony and scroll design on the central medallion (Photograph 55). The piece has a reign mark of either 'Xuande' (宣德) or 'Zhengde' (正德), although it lacks the first character for a clear identification, but from the light cobalt colour and its design it is probably dated to the early half of the 16th century, possibly the Zhengde period (1506-1521). This type of large bowl can be found among the heirloom collections in Lisbon, but it was no longer produced by the latter half of the 16th century.

G) A large plate with wave and fish design on the interior and scroll and conch motif on the exterior. The design is again drawn with light cobalt which is characteristic of this period. It is thickly potted, a feature totally absent from late 16th-century ware (Photograph 56). Thus it is probably dated somewhere in the first half of the 16th century at the latest. From the exterior design it may be dated to the early decades of the century. No identical piece is found from any datable shipwreck or land sites, although there are stylistic links to type C above.

4-3. The mid 16th century to early Wanli period

I) Plate with scroll design or emblem on a flat rim. It is a plate that can be commonly found among the collections of Fundação Almeida and Anastácio Gonçalves in Lisbon

 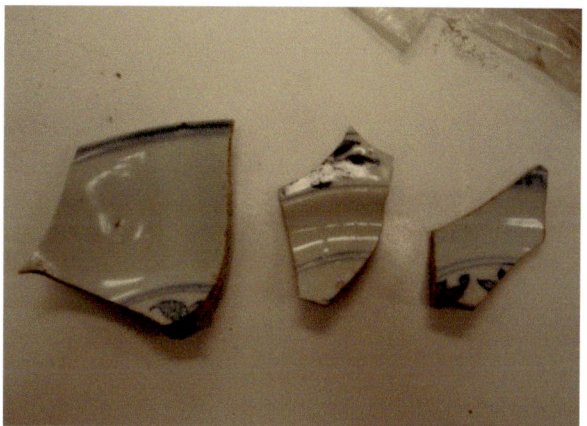

Photograph 53: Blue-and-white bowl with floral spray design excavated from Alfama (left) (© Museu da Cidade).

Photograph 54: Chinese ceramics with cross mark on the base excavated from Alfama (© Museu da Cidade)

Photograph 55: Chinese ceramics with reign marks on the base excavated from Alfama (© Museu da Cidade).

Photograph 56: Jingdezhen blue-and-white large dish with wave design on the interior and scroll design on the exterior excavated from Alfama (© Museu da Cidade).

and can be dated to the mid 16th century. The central medallion is surrounded by a repeated scroll or *ruyi* motif, painted with bright cobalt. In most cases they have stylized designs drawn on the flat rim and the cavettos are left blank or decorated with ribbons. The design can be recognized as a prototype of the *kraak* type, which would later become extremely popular in all European markets (Photograph 57).

J) Bowl with circular motif on the exterior wall. This circular motif is frequently found among the pieces dated from the Jiajing period to possibly early Wanli (1572-1620) at the latest. A good example of this type is present among the cargoes of the wreck of the *San Sebastian* off Mozambique which sunk in the late Jiajing period. The bowl has a stright standing rim with circular motifs, which in some cases are stylized dragons or flowers drawn on the exterior wall (Photograph 58).

K) Bowl with floral design, or bird on a twig motif on the exterior wall (Photograph 59). This is a bowl with a straight standing rim and slightly rounded wall. The exterior wall is always decorated in rather dark cobalt with a floral or bird motif. From the colour of the cobalt and brushwork it is of late Jiajing period production to either Longqing (1567-1572), or very early Wanli, sometime in the 1560s and 1570s. A similar type is present among the pieces excavated in Seville and Mexico City, and this helps provide a date of c. 1570, after the beginning of the Manila galleon trade in 1565.

4-4. The third quarter of the 16th century

L) Small bowl with the raised base style generally known as *wantouxin* (万頭心) with a combination of flower, fruit, phoenix and human figure motif in the centre (Photograph 60). This is a very common type found in almost all contemporary port sites in Southeast Asia as well as Japan (i.e. the excavations at Ichijodani). The diameter of the foot ring is small and the foot is normally slanted inwards. The wall is straight to the rim and frequently left without decorations.

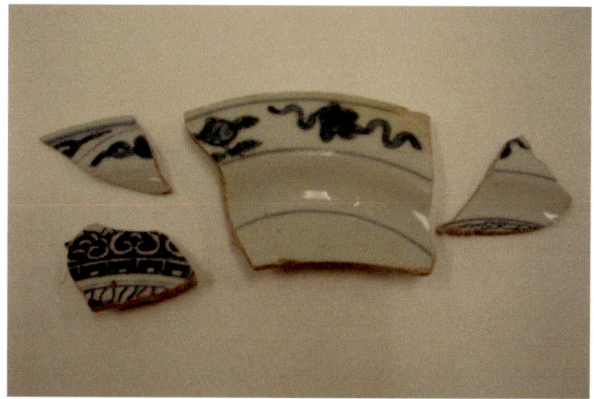

Photograph 57: Jingdezhen blue-and-white large plates excavated from Alfama(© Museu da Cidade).

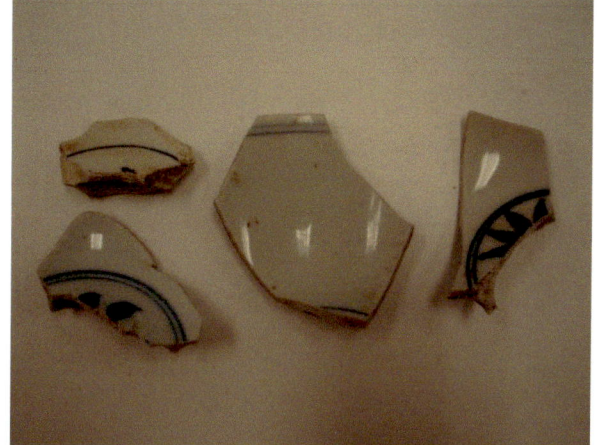

Photograph 58: Jingdezhen blue-and-white sherd of small bowl with circular motif on the exterior excavated from Alfama(© Museu da Cidade).

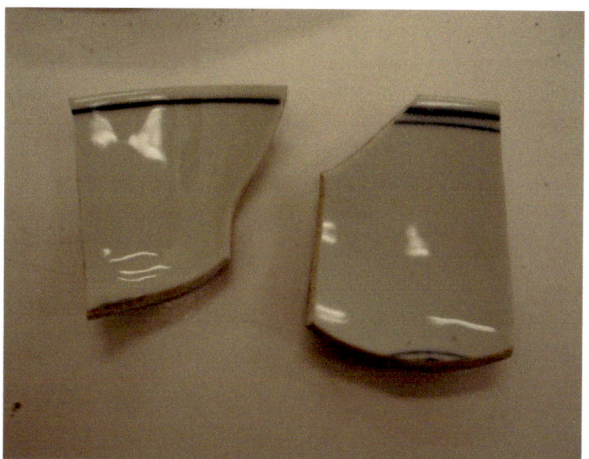

Photograph 59: Jingdezhen blue-and-white sherds of bowls excavated from Alfama(© Museu da Cidade).

M) Plate with eight Buddhist emblems on the flat rim and normally with a central phoenix (Photograph 61, below left). This is a type abundantly found among the *Golden Hind* items and the Zócalo area of Mexico City, and was probably produced and exported in large quantities from China during this period to New Spain and other markets. It is normally moulded and then the foot ring retouched and carved out. This type is rarely found at Alfama site, which may give an idea to the dating of this excavation site.

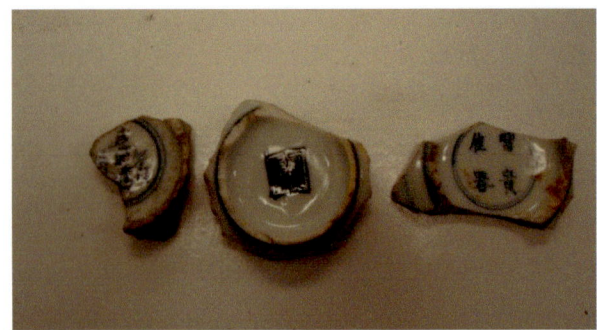

Photograph 60: Jingdezhen blue-and-white sherds of bowls generally known as *wantouxin* excavated from Alfama (© Museu da Cidade).

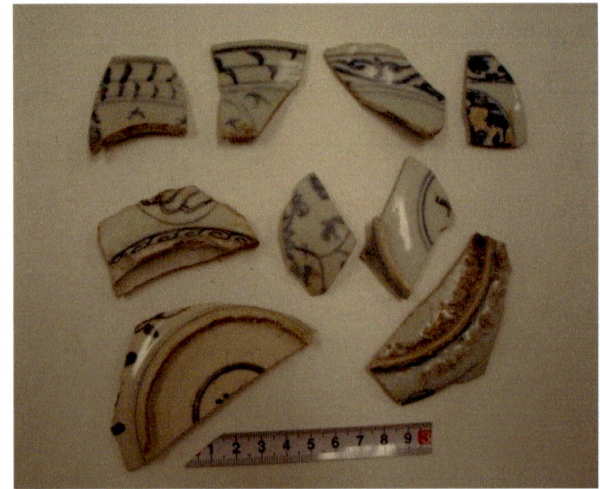

Photograph 61: Jingdezhen blue-and-white plate with phoenix motif on the central medallion excavated from Alfama (© Museu da Cidade).

The ceramic materials classified here were produced and traded throughout the 16th century. Most of the ceramics excavated from the Alfama site were concentrated within this century and very few pieces from the 17th and 18th centuries have been found. Ceramics from the first half of the 16th century are most abundant and varied and thus define more or less the dating of this archaeological complex.

The ceramics of the early 16th century found from this site are among the earliest types that exist in Europe as traded materials in large quantities. Chinese ceramics of the 16th century are only found in abundance in Portugal as the merchants and others here were the only direct importers of Asian goods in Europe. However, early 16th century pieces are even rare in Portugal and these ceramics can be very important in terms of the study of trade with Asia in this period.

5. The porcelain trade from Lisbon to Amsterdam

In relation to Chinese porcelain, there is an apocryphal story that the first Dutch person to own cabinets of real porcelain was Amalia Hendrik, wife of Frederik Hendrik (1584-1647).[183] With the founding of the Dutch East India Company in 1602, the Dutch were able to acquire luxury Chinese porcelain directly from Asia. From the first half of the 17th century onwards Chinese porcelain began to be imported in large quantities by the Dutch, as can be seen from several shipwrecks, such as the *Witte Leeuw*, which sank off Santa Helena in 1613, and the *Hatcher* cargoes (dated c. 1643-1646), lost in the South China Sea.

However, before the founding of the Dutch East India Company, Dutch merchants probably acquired Chinese porcelain from Lisbon. From the medieval period, Lisbon exported wine, salt, cereal and dry fruits to the north. After the fall of Antwerp, Amsterdam was the main port of entry for these products.[184]

Amsterdam had already gained its importance as a port-city by the mid 16th century, although the merchants residing in Amsterdam were it seems all originally from

---

[183] Jan van Campen, 'Asian Ceramics in the Netherlands', *Asian Art & Culture*, Amsterdam, 2003, p. 42.
[184] Catia Antunes, *Lisboa e Amsterdao 1640-1705: Um caso de Globalizacao na Historia Moderna*, Livros Horizonte, Lisbon, 2009, pp. 32-50.

the surrounding areas or from Amsterdam itself. Trade was mostly carried out by the reception of ships from the Baltic countries.[185] Before the rise of Amsterdam, Antwerp was the more prosperous city as a trading port and its fall caused the creation of a merchant diaspora, most of them leaving for other centres. It was when these merchants left Antwerp for locations such as Amsterdam that they began to use their trade networks to access most of the major cities in Europe. At the same time, Amsterdam merchants began to provide capital to the merchants from Antwerp and there was a gradual development of trade routes to France, Spain and Portugal, and ships began to operate regularly between these countries.

The city of Amsterdam began extensive excavation works in the 1970s and the project is still ongoing. Early ceramic pieces have been found in strata ranging from 1592-1596, coinciding with the finds from Drake's Bay (*Golden Hind*) and the *San Augustin* pieces dated to 1597.[186] These items were most probably imported via Lisbon, but the later pieces – such as the two plates from the 1640s that match the *Hatcher* cargoes – were directly shipped from Banten. One interesting result of the excavation, in art historical terms, is that it was always said that the Europeans avoided inferior quality ceramics from Zhangzhou province, although the the present author has noted at least three pieces from this province from the 57 Chinese ceramics found so far. Of course, these excavated pieces represent only one limited category of the Chinese ceramics imported to Amsterdam, as the present author has only focused on early pieces that relate to the Atlantic coastal trade with Lisbon. Later pieces, such as 18th-century assemblages and Japanese Hizen wares, are in famous museum collections. Thai storage jars of the 17th century were also found used as containers, although their contents are unknown.

The majority of the ceramic materials excavated from Alfama in Lisbon consists of early to mid 16th century pieces. Ceramics dated from the Zhengde to the Jiajing periods are varied, although their volumes and types decrease towards the end of the century. Official trade between Portugal and China (based on Macao) was not permitted in the early to mid 16th century and it may be concluded that ceramics in this period were traded in Malacca and also bought and sold (as contraband) in the regions near southern China. The fact that identical types of ceramics with cross mark were found in Lisbon, Hong Kong and Macau indicates that these pieces exported towards the Portuguese markets were traded privately in the Guangdong area. If prices of Chinese products doubled when reaching Malacca, as the document shows, it is probable that the Portuguese preferred to buy trade goods in China as contraband wherever they could at a better price, rather than trading through legitimate channels in Malacca. Early 16th-century Chinese ceramics in Europe are normally understood to be bespoke pieces bearing family crests and other European motifs. Blue-and-white porcelain decorated with the escudo of King Manuel I is well known among heirloom pieces (see Chapter I). These were especially ordered by wealthy individuals as special gifts and it is not known whether trade in large quantities carried out in this period. However the excavated materials from Alfama are obviously traded objects as they are found in some number and in many instances with typical Jingdezhen designs of the period rather than special Occidental motifs for personal orders. We know that from the early stages of Portuguese expansion towards Asia in the early half of 16th century that they were already starting trade in ceramics via Malacca and southern China, in most cases as contraband, and transporting them directly to Lisbon. This distribution of Chinese ceramics in the 16th century by the Portuguese continued for over 100 years, which was long enough to engender the boom in Chinese ceramics in the Netherlands and other European markets in the 17th and 18th centuries.

The Chinese ceramics found in Bayona and Vigo represent one aspect of the Atlantic coastal trade's history during the late 16th and early 17th centuries. The excavated ceramics from these two sites were probably traded by the Portuguese and brought to Lisbon. From there they were further transported to Oporto or Viana do Castelo, and then, clandestinely or by normal trade, imported to the Galician region with other products such as sugar and salt. These were dealt in by either the Portuguese or Galician merchants who frequently went back and forth between the two adjacent countries. The fact that Galicia had trade links with England and the Netherlands, especially movements of large quantities of olive oil, wool and linen, may leave open the possibility that these ceramics were transported by English or Dutch merchants. Sugar was distributed within the region or passed through on its way to Asturias and the Basque region where there was a larger market. Moreover, many ships loaded with sugar from Madeira and Brazil headed directly to Galicia clandestinely.

In any event, these ceramics were most probably brought from Portugal, and not by the Manila galleon trade, since many of the pieces coincide with those in Portuguese collections. The small quantities of excavated ceramics indicate that Chinese ceramics were not a trade regular item between Portugal and Galicia and some pieces may have been brought in by smuggling. Furthermore, the dating of the pieces that are concentrated in the mid to third quarter of the 16th century supports the above theory that these items were more probably brought in via Portugal than via the Mexico–Seville route, a factor related to the sugar trade as it grew in the mid 16th century.

---

[185] Katsumi Fukazawa (ed.), *Kokusaishougyo Kindai Yoroppa no Tankyu*, Vol. 9, Minelvashobo, Kyoto, 2002, p. 55.
[186] The information comes from personal interviews with an archaeologist in charge of the Amsterdam City excavation project.

Even though the Chinese ceramics found along the Atlantic coast are out of context with the Manila galleon trade route, it is important to consider the connections of the merchants active at the time. These indicate that Asian goods were widely distributed to America along the Pacific trade links, and also distributed throughout Europe mainly via Lisbon. Chinese ceramics reaching Lisbon were greatly appreciated in Portugal and were further transported along the Atlantic coast towards northern Spain and the Netherlands (Amsterdam) and from there connected to the Baltic trade – which grows in importance in the 17th and 18th centuries.

This Atlantic coastal trade was also undertaken by some of the Portuguese *conversos* who had been dispersed and who had connections between Lisbon, Galicia and Amsterdam.

6. Conclusion

The Manila galleon trade, which continued for three centuries, was not just about the trade of silks for silver, but a dynamic force for change that brought goods, people and cultural exchanges from distant regions. The first impetus for this was generated by the Portuguese: they already knew of Asian trade and the region fairly well, and, of course, was also the Iberian country adjacent to Spain. Without Portuguese support the Manila galleon trade might not have flourished so readily after the settlement in by Miguel López de Legazpi in Cebu.

From its beginnings in the 15th century Asian trade was quickly exploited by the Iberian powers from the early 16th century, first by the Portuguese after conquering Goa and then Malacca. Unofficial trade then began with the Chinese and the Japanese. Portuguese connections with Chinese merchants in southern China, having benefited from access to Japanese silver from the first half of 16th century, developed into links with the complex network of Southeast Asian commercial products based on settlements in Malacca and other Asian ports. Trade was actively carried out by private merchants. Among these skilled and wealthy entrepreneurs there were some (if not many) Jewish *conversos* who had fled the Portuguese Inquisition and settled in Asia and who had a substantial family networks stretching from Asia to Europe – especially in Amsterdam and Hamburg. Although it is not clear how large their communities were in Goa, Malacca and Macao, there was a well established Jewish community, with a synagogue, in Malacca in 1547, according to Jesuit sources.[187]

Some *conversos* escaped the Inquisition in the other direction, crossing the Atlantic and settling in the Americas. 'New Spain' was a region offering great business opportunities, especially in Mexico City, La Puebla, Guadalajara, Oaxaca, San Luis and Veracruz. Peru was another Spanish colony with large amounts of silver, and later Brazil proved a magnet for *conversos*, who could link directly with their families and friends, especially in Amsterdam.

In terms of trade in the Asian region, investments by private Portuguese traders, such as the above-mentioned Landeiro family, based in Macao, must have led to a financial power-base able to support the sending of private ships to Japan, Siam, and other Asian and Far Eastern lands. When the Spanish conquered the Philippines, according to some primary sources mentioned in Chapter I, they were not confident of profiting from these scattered islands. There were so few people to establish a firm colony and begin ruling the region from the outset. Furthermore, their knowledge of Asian trade and the geography were very limited and the governors waited desperately for more military and financial support from New Spain, rather than settling down to finding the means to support themselves by local trade. In this situation it was probably the financial and other direct support of private Portuguese merchants that led to the inception of the Manila galleon trade from 1565. Although the plaintive missives for help from Legazpi on one hand, and the galleons sent to Acapulco on the other, seem to be contradictory, the Jingdezhen fragments dated to the mid 16th century, some of which are identical to finds among the heirloom pieces in Lisbon, indicate the distinct possibility that the Portuguese were the agents who actually pushed the business forward. The exports of fairly large quantities of silk that appear in the registers of cargoes also indicate that the Portuguese had an important role in trans-Pacific trade and the supply of Asian goods via Macao. While in relation to Chinese merchants in Manila they were also an important factor in keeping Manila and the Philippines operating as way stations in Asia, it has been explained earlier that the Portuguese were more directly connected to the Manila galleon trade, aware of the demand for various products by the American and European markets. The southern Chinese merchants who came to Manila and played an important role in trade were probably the major suppliers of silk and other goods to be consumed within Manila or other cities in the Philippines, whereas the Portuguese seemed to focus on goods destined for New Spain. It has already been pointed out by Charles Boxer that some of the Portuguese merchants in Macao were agents of the Mexican merchants who had invested heavily, especially in Chinese silk.[188]

Especially from about the third quarter of the 16th century, the Manila galleon trade prospered for nearly 100 years. This can be seen by the quantities of Chinese ceramics excavated in Mexico (see Chapter III) and

---

[187] Walter J. Fischel, 'New Sources for the History of the Jewish Diaspora in Asia in the 16th century', *The Jewish Quarterly Review*, Vol. 40, No.4, University of Pennsylvania Press, 1950, pp. 394-395.

[188] Charles R. Boxer, *op. cit*, p. 247.

also by the studies of Luisa Schell Hoberman.[189] Among the large investors in the Manila galleon trade there were the *conversos*, such as the previously mentioned Simón Váez de Sevilla, who migrated from the Iberian Peninsula and had his family network all over Europe and even possibly in Asia. Bartholome Landeiro, as we have seen, was another *converso* who migrated to Macao and had extended family networks in Europe and New Spain.[190] The Millán family was another powerful *converso* agency with family members in Antwerp, Hamburg, Amsterdam, Livorno, Venice, Orinda (Brazil), Cabo Verde, Mexico City, Acapulco, Goa, Manila and Macao;[191] Félix Millán of Mexico City appears as the merchant importing Asian goods in 1640.[192]

It is very probable that these Portuguese *conversos* operating between New Spain, Macao and Manila had a certain influence over the Manila galleon trade, and Macao *conversos* worked as agents of Portuguese *conversos* in Mexico who were connected mostly by family ties. One example is Sebastian Váez Acevedo, a captain general and a merchant in New Spain; he had a brother Antonio Váez Acevedo who was a mayor of Panpanga in the Philippines. Antonio was not a merchant although he might have sent goods on behalf of this brother.

To what extent these Portuguese *conversos* had an influence on the Manila galleon trade is yet unknown but at the very least their worldwide connection, especially their trade network in Asia, was an important element for the inception and growth of this trans-Pacific trade. The recession of the mid 17th century had many factors underpinning behind: a Chinese civil war succeeded by the conflict with Koxinga in Taiwan; Dutch competition in Asia; and the decline in the Indian labour force working in the American silver mines, which led to a decline in output. These were all factors contributing to the slow down in activity and concentrated in the 1640s and 1650s. Adding to this, the Mexican Inquisition, which afflicted the city in the 1640s onwards, may have disrupted the networks of the *conversos* in terms of trade in all their markets, temporarily influencing the export of Asian goods to New Spain.

The important, shared factor for these Portuguese *conversos* in Asia and in America was that they both had family members in Antwerp, Amsterdam and Hamburg. This means that it was not a coincidence that flows of Asian goods shifted towards the Atlantic and further to the Baltic Sea. The movements of items from Lisbon northwards to Amsterdam was an obvious consequence given the existing networks of the *conversos*. This Atlantic coastal trade north from Lisbon is not directly related to the Manila galleon trade, but by tracing fragments of Chinese ceramics, which were one of the main products exported from Asia to the West, it becomes clear how the trade network extended into Europe and beyond.

Further study on how Portuguese *conversos* in America and Asia were connected, and how Manila and Macao merchants, who were Portuguese *conversos*, actually carried out this trade as agents, is required. It is necessary, therefore, to trace the family trees of those individuals dispersed to Macao, Malacca, Goa, Lisbon and Evora, comparing and referencing to Inquisition documents and the cargo registers in Mexico. Their diaspora and activities in Asia, and integration within local societies, will also help us understand the structure of trade in Asia, together with further archaeological research on Chinese ceramics in Macao and Manila.

---

[189] Luisa Schell Hoberman, *op. cit*, pp. 39-40.
[190] Lucio de Sousa, *op. cit*, pp. 229-231.
[191] Lucio de Sousa, 16-17 Seiki no Portugal jin ni yoru Asia Dorei boueki-Victoria Diaz Aru Chugokujin joseidorei wo otte, *Namban, Koumo, Toujin: 16-17seikino Higasi Asia Kaiiki*, Shibunkaku Shuppan, Tokyo, 2013, p. 258.
[192] AGN Indiferente Virreinal, caja-exp.: 4976-006.Fols.1r-64v

Glossary

Jingdezhen (景德鎮**)**

Located in Jianxi province, ceramic production is said to have begun as early as the Han Dynasty (206 BC – AD 220). In the Song Dynasty the area began to produce high-quality *qingbai* wares (青白磁), which are tinted slightly light blue. In the Yuan Dynasty popularity grew rapidly with the production of blue-and-white wares and these were exported all over the world through the Yuan, Ming, and Qing dynasties.

Zhangzhou (漳洲**)**

Located in Fujian province, the exact period of commencement of ceramic production here is unknown, although low-fired rough ceramics, with grey to tinted blue cobalt surfaces, began to be exported from the 16th century. Zhangzhou wares are generally called 'Swatow ware' from the name of the loading port from where they were exported, mostly to countries around Asia.

Porcelain (磁器**)**

A ceramic material containing kaolin and fired to temperatures of 1200 to 1400 °C. The body becomes almost translucent white with a thin glass-like texture.

Ceramic (陶器**)**

Ceramic is a general term for inorganic, non-metallic solid wares made from clay and heated to become solid. The term includes low-fired pottery and also porcelain (see above).

Blue-and-white (青花**)**

Ceramics with decoration using oxidized cobalt pigment and fired at around 1300 °C.

Ruyi head (如意頭**)**

Ruyi, meaning 'may the wish be granted' is depicted usually as an inverted, heart-like shape.

Lingzhi (霊芝**)**

The 'sacred fungus', a Taoist symbol of longevity.

Taihu rockery (太湖**)**

Taihu rocks originate from the eponymous region near Suzhou (China). Historically they were much loved by many emperors for their shapes, caused by hydraulic erosion.

Eight Taoist Symbols (八宝文**)**

These eight traditional Taoist motifs are: fan, bamboo tube and rods, sword, castanets, double gourd, flute, flower-basket and lotus. The motifs are attributed to the Eight Immortals and often feature on Chinese ceramic decoration.

Chinese Dynasties and Periods

Tang Dynasty (唐**): 618-907**

Liao Dynasty (遼**): 907-1125**

Song Dynasty (宋**): 960-1279**

Northern Song Dynasty (北宋**): 960-1127**

Southern Song Dynasty (南宋**): 1127-1279**

Yuan Dynasty (元**): 1279-1368**

Ming Dynasty (明**)**

Hongwu (洪武): 1368-1398
Jianwen (建文): 1399-1402
Yongle (永楽): 1403-1424
Honxi (洪熙): 1425
Xuande (宣徳): 1426-1435
Zhengtong (正統): 1436-1449
Jingtai (景泰): 1450-1457
Tianshun (天順): 1457-1464
Chenghua (成化): 1465-1487
Hongzhi (弘治): 1488-1505
Zhengde (正徳): 1506-1521
Jiajing (嘉靖): 1522-1566
Longqing (隆慶): 1567-1572
Wanli (万歴): 1573-1620
Taichang (泰昌): 1620
Tianqi (天啓): 1621-1627
Chongzhen (崇禎): 1628-1644

Qing (清**) Dynasty**

Shunzhi (順治): 1644-1661
Kangxi (康熙): 1662-1723
Yongzheng (雍正): 1723-1735
Qianlong (乾隆): 1736-1795
Jiaqing (嘉慶): 1796-1820
Daoguang (道光): 1821-1850
Xianfeng (咸豊): 1851-1861
Tongzhi (同治): 1862-1874
Guangxu (光諸): 1875-1908
Xuantong (宣統): 1909-1911

Bibliography

Unpublished Materials

ARCHIVES
AGI: Archivo General de Indias (Seville)
AHN: Archivo Histórico Nacional (Madrid)
AGN: Archivo General de Nación (Mexico Districtro Federal)
Archivo Histórico SJ de Cataluña (Barcelona)
AGI Filipinas, 6, R.1, N.1, *Carta de Legazpi avisando su llegada y establecimiento*
AGI Filipinas, 6, R.1, N.5, *Carta de Legazpi sobre descubrimientos realizados y armas*
AGI Filipinas, 6, R.1, N.7, *Carta de Legazpi sobre falta de Socorro y descubrimientos*
AGI Filipinas, 6, R.1, N.8, *Carta de Legazpi pidiendo Socorro a Nueva España*
AGI Filipinas, 6, R.1, N.12, *Carta de Legazpi sobre descubrimiento en el norte*
AGI Filipinas, 6, R.2, N.16, *Carta de Guido de Lavezaris sobre los esclavos de Filipinas*
AGI Filipinas, 27, N.30, *Suplica VM le haga Vuestra merced proveyendo en todo lo que convenga al servicio de VM y a la conservación y aumento deste Reyno*
AGI Filipinas, 27, N.148, *Petición del Cabildo secular de Manila sobre Parián de Sangleyes*
AGI Filipinas, 27, N.156, *Varios documentos sobre proteger a comerciantes de Filipinas*
AGI Filipinas, 28, N.131, *Memoria y Lista de los oficios que tiene y ejercen los Sangleyes Cristianos que reciben y moran intra y extramuros de la Ciudad de Manila*
AGI Filipinas, 28, N.160, *Petición del Cabildo secular de Manila sobre carga de naos*
AGI Filipinas, 29, N.10, *Cuentas Sobre China*
AGI Filipinas, 29, N.57, *Carta de Francisco de las Misas sobre varios asuntos de Filipinas, comercio, salario*
AGI Filipinas, 35, N.3 *Carta de Juan Nuñez sobre su pobreza*
AGI Filipinas, 34, N.35, *Mercancías embargados a Francisco de Sande*
AGI Filipinas, 35, N.82, *Testimonio del numero de Sangleyes que entran en Manila*
AGI Filipinas, 41, N.16, *Petición de la ciudad de Manila sobre comercio de Portugueses*
AGI Filipinas, 64, N.1, *Registros de champanes y pataches llegados a Manila*
AGI Filipinas, 66, *Cartas y testimonios de autos obrados en Acapulco México y Filipinas en razon de decubrir los bienes del gobernador de dichas yslas don de Juan de Vargas Hurtado y su cuñado Francisco Guerrero de Ardilla año de 1685 a 1688*
AGI Filipinas, 70, *Autos sobre barcos de Macao*
AGI Filipinas, 74, N.137, *Carta de Juan López sobre bula de comercio de eclesiásticas*
AGI Filipinas, 77, N.6, *Carta del cabildo eclesiástico de Manila sobre comercio de Filipinas con Nueva España*
AGI Filipinas, 82, N.1, *Copia de Real Cédula a la Audiencia sobre comercio de Portugueses*
AGI Filipinas, 84, N.3, *Memoria de lo que fray Diego de Herrera ha de tratar en la corte*
AGI Filipinas, 85, N.34, *Carta de Dominico Diego Aduarte sobre comercio*
AGI Filipinas, 340, L.3, *Orden sobre esclavos de pasajeros en las naos de Filipinas*
AGI Indiferente, 745/1598-1599/ Consulta Indiferente General, N.163 29/11/1598, *Sobre la convivencia de que el Consejo de Portugal despache orden para que los navíos que desde Perú fueron a las Indias vuelvan a Portugal*
AGI Contaduria, 1196/1565-1576, *Caja de Filipinas, Cuentas de Real Hacienda*
AGI Contratación, 1795-1802, *Registros de Venida del navíos*
AGI Contratación, 5262A, N.62, *Expediente de información y licencia de pasajero a indias de Antonio Rodríguez Arias, mercader, vecino de Sevilla, hijo de Simón Rodríguez y Catalina Márquez, a La Habana*
AGI México, 19-82, fol.3, *Carta de Martín Enriquez a SM*
AGI Patronato, 24, R.20, *Titulo de Ciudad de Manila*
AHN Inquisición, 4812, Exp.16, *Cuentas de los bienes secuestrados a Juan Duarte de Espinosa*
AHN Inquisición, 4812, Exp.5, *Cuentas sobre los bienes de Thomas Nuñez*
AHN Inquisición, 4812, Exp.3, *Cuentas de los bienes secuestrados a Simón Fernández de Torres, Simón Suárez de Espinosa y otros*
AHN Inquisición, 4812, Exp.4, *Cuenta de los bienes secuestrados y confiscados a Tomás Treviño de Sobremonte y otros*
AGN Archivo Historico de Hacienda, leg. 268, exp.196
AGN GD8, Archivo Histórico de Hacienda, Leg. 472, exp. 126, OFICIALES REALES
AGN Archivo Histórico de Hacienda, leg.1291
AGN Caja real Filipinas, 1251, *Los bienes sin registrado en Cavite 1697*
AGN General de Parte, Vol. 4, exp. 339, fs. 98, *Nao de Macan 1590*
AGN Indiferente Virreinal, caja-exp. 6441, exp. 092, 1640
AGN Indiferente Virreinal, caja-exp. 4976-0006, *Mercadurias*
AGN Indiferente Virreinal, caja-exp. 4259-012
AGN GD64 JESUITAS, Año: 1581-1645, Vol. IV, 50
AGN GD100 Reales Cedulas, Vol. 11, Exp. 451, fs. 317

Archivo de los Jesuitas (Barcelona), EI. C-7/2/7 *Cartas y relaciones de las oficiales de Filipinas sobre la llegada y cerco de los Portugueses*, Doc. Nº 43

Keresey, Déborah Oropeza, *Los "indios chinos" en la Nueva España; la inmigración de la nao de China,1565-1700*, El Colegio de México, Ph.D. dissertation, 2009.

Rivas Castro, Francisco. Excavaciones Recientes en la Calle de Licenciado Verdad No.3, Centro de la Ciudad de México. (Presented at the Sociedad Mexicanas de Antropología, Universidad Veracruzana, 1992, unpublished).

Published Materials

Aguado de los Reyes, Jesus. *Riqueza y Sociedad en la Sevilla del Siglo XVII*, Fundacion Fondo de Cultura de Sevilla y Universidad de Sevilla,1994.

Ahmad, Afzal. *Portuguese Trade and Socio- Economic Changes On the Western Coast of India 1600-1663*, DK Publishers. Delhi, 2000.

Aoyagi, Yoji. Koeki no jidai 9~16 seiki no Philippine, *Jochi Asia Gaku*, Vol.10, Tokyo, 1992, pp. 144-176.

Alonso, Diego Oliva (coordinador). *Retrato del Conde de Altamira*, Junta de Andalucia, Novograf, Sevilla, 2005.

Anitua, Fernando Tabar de. *Cerámicas de China y Japón en el Museo Nacional de Artes Decorativas*, Ministro de Cultura, 1983,

Antunes, Catia. *Lisboa e Amsterdao 1640-1705: Um caso de Globalizacao na Historia Moderna*, Livros Horizonte, Lisbon, 2009.

Azogue, Araceli Rodríguez Y Luengo, Vicente Aycart. *San Juan de Acre: La Historia Recuperada de un Barrio de Sevilla*, Emvisesa, Sevilla, 2007.

Barreto, Luís Felipe (ed.). *Macau During the Ming Dynasty*, Centro Cientifico e Cultural de Macau, I.P, 2009.

Bjork, Katherine. 'The link that kept the Philippines Spanish, Mexican Merchant Interest and the Manila Trade, 1571-1815', *Journal of World History*, Hawaii University Press, Vol. 9, No.1, Spring 1998, pp. 25-30.

Borao, José Eugenio. *Spaniards in Taiwan*, SIT, Taipei, 2001.

Boxer, Charles R. *O Grande Navio de Amacau*, Fundación Oriente y Centro de Estudios Marítimas de Macao, Macao, 1960.

Boxer, Charles R. *The Fidalgos in the Far East, 1550-1770*, Oxford University Press, London, 1968.

Boxer, Charles R. *Macao 300 years ago*, Fundação Oriente, Macao, Lisboa, 1942.

Brown, Roxanna M. *The Ming Gap and Shipwreck Ceramics in Southeast Asia; Towards a Chronology of Thai Trade Ware*, The Siam Society under Royal Patronage, Bangkok, 2009.

Cambra, Rosario Huarte y Muñoz, Pilar Some. 'Ceramica Contemporanea en el Cuartel del Carmen (Sevilla)'*, SPAL* 4 (1995), pp. 229-247.

Calo Lourido, Francisco (cood.). 'Do Porto a Pontevedra (XIII-XVII): Os Tratos Maritimos no Noroeste Peninsular. Do Porto à Galiza e a Pontevedra', *Actas do Simposio de Historia Marítima do Século XII ao XVI*, Pontevedra, Graf, Duher, 2003.

Caramés Moreira, Vicente and Cobo Rodríguez, Fátima. 'Porcelana chinesa da dinastía Ming Procedente do Parque da Palma de Baiona'. Castrelos, nº.13, Vigo. Museo Municipal de Vigo Quiñones de León, 2008, pp. 96-106.

Careri, Giovanni Francisco Gemelli. *Viaje a la Nueva España: Estudio Preliminar, Traducción y notas de Francisca Perujo*, Uiversidad Nacional Autóoma de México, 1976.

Carletti, Francisco. *Mi Viaje Alrededor del Mundo (1594-1606)*, Editorial Noray, Barcelona, 2006.

Campen, Jan van. 'Asian Ceramics in the Netherlands', *Asian Art & Culture*, Amsterdam, 2003, p. 42.

Cervantes, Gonzalo Lopez. 'Porcelana Oriental en la Nueva España', *Anales de INAH (Instituto Nacional Arqueológico e Histórico)*, México, 1977.

Colín, Ostwald Sales.'Las Cargazones del Galeón de la Carrera de Poniente', *Revista de Histórica Económica*, 2008, Otoño-invierno, pp. 645-646.

Chaunu, Pierre. *Les Philippines et le Pacifique del Ibérics (XVI, XVII, XVIII siécles)*, S.E.V.P.E.N, París, 1960, pp. 204-206.

Chia, Lucille. 'The Butcher, the Baker, and the Carpenter: Chinese Sojourners in the Spanish Philippines and Their Impact on Southern Fujian (Sixteenth-Eighteenth Centuries)', *Journal of the Economic and Social History of the Orient*, Vol. 49, No.4, Maritime Diasporas in the Indian Ocean and East and Southeast Asia (960-1775), 2006, pp. 509-534.

Crick, Monique. 'The San Diego Galleon, 14 December 1600, a Dating for Swatow Porcelains', *Oriental Art*, Vol. 46, No.3, 2000, pp. 22-31.

Cushner, Nicholas. *Spain in the Philippines*, Ateneo Tuttle, 1971.

Flanderín, Jean-Louis. 'El Proceso de cambio en la sociedad de los siglos XVI-XVII', *La Distinción A Través del Gusto en Historia de la Vida Privada de Philip Arles y Georges Duby. (Tomo V)*, Editorial Taurus, Madrid, 1992.

Fuentes, Lutgardo García. *El Comercio Español con América (1650-1700)*, Escuela de Estudios Hispano-Americanos, Sevilla, 1980.

Fukazawa, Katsumi (ed.). *Kokusaishougyo* Kindai *Yoroppa no Tankyu*, Vol. 9, Minelvashobo, Kyoto, 2002.

Gage, Thomas. *Nuevo Reconocimiento de las Indias Occidentales*, 1ª edición, 1648, Fondo Cultura Económica, México D.F., 1982.

García-Abásolo, Antonio. La Audiencia de Manila y los Chinos de Filipinas. *Casos de Integración el Delito*, Universidad Nacional Autónoma México.

García-Abásolo, Antonio La llegada de los españoles a Extremo Oriente y la colonizacion de Filipinas, Gran Historia Universa, Vol. XXVII, *Descubrimiento y Conquista de America, Club Internacional del Libro*, Madrid, 1982.

Gil, Juan. Los Chinos en Manila SiglosXVI y XVII, Centro Científico Cultural de Macau, I.P.Lisboa,2011.

Gonzales Muñoz, Mª del Carmen. 'Vigo y Su Comarca en los Siglos XVI y XVII'. *Vigo en Su Historia*, Artes Gráficas Galicia, Vigo, 1979, pp. 153-276.

Gonzales Muñoz, Mª del Carmen. *Galicia en 1571: Poblacion y Economia*, Edicios do Castro Serie Liminar historia, A Coruna, 1982.

Hirayama, Atsuko. *Spain Teikoku to Chuka Teikoku no Geko, 16, 17 seikino Manila*, Hoseidaigakushuppan, Tokyo, 2012.

Hirosue, Masashi. *Tonan Asia no Koushi Sekai*, Chiikishakai no Keisei to Sekai Chitsujo, Iwanami Shoten, Tokyo, 2004.

Hoberman, Schell Louisa. *Mexico's Merchant Elite. 1590-1660, Silver, State and Society*, Durham and London, Duke University Press, 1991.

Hoberman, Schell Louisa.'Merchants in Seventeenth-century Mexico City: a preliminary portrait', *Hispanic American Historical Review*, 57: 3, 1977.

Hoberman, Schell Louisa and Socolow, Susan (eds.), *Cities and Society in Colonial Latin America*. Alberquerque: University of New Mexico Press, 1986.

Haring, Clarence H. *Comercio y Navegacion entre España y las Indias*, Fondo de Cultura Económica, México, 1939.

Homer, H. and Smith, Robert S. 'Chinese in Mexico City in 1635', *The Far Eastern Quarterly*, Vol.1, No.4, Aug. 1942, pp. 387-389.

Huang, Wei and Huang, Qing, Guangdong Taishangdao Huawanpinyishi Chutu Ciji ji Xiangxiang guan Wenti, Wenwu Di wu qi, Beijing, 2007, pp. 78-88.

Israel, Jonathan I. *Razas, Clases Sociales y Vida Política en el México Colonial 1610-1670*, Fondo de Cultura Económica, México, 1981.

Impey, Olivier R. *Cerámica do Extremo Oriente*, Casa Museo Guerra Junqueiro, Porto, 1992.

Jeannin Pierre. *Merchants of the Sixteenth Century*. New York: Harper and Row, 1972.

Karashima, Noboru. *In Search of Chinese Ceramic-sherds in South India and Sri Lanka*, Taisho University Press, Tokyo, 2004, pp. 4-62.

Kimura, Masahiro. *La Revolución de los Precios en el Pacífico:1600-1650*, Universidad Nacional Autónoma de México, 1987.

Kitahara, Michio. *Portugal no Shokuminchi Keisei to Nihonjin Dorei*, Kodensha, Tokyo, 2013.

Kuwayama, George, *Chinese Ceramics in Colonial Mexico*, Hawaii University Press, Hawaii, 1997.

Lach, Donald F. *Asia in the Making of Europe Vol. I: The Century of Discovery*, The University of Chicago Press, Chicago and London, 1965.

Lopez-Cano, María del Pilar Martínez. 'Los Mercaderes de la Ciudad de México en el S. XVI y el comercio con el exterior', *Revista Complutense de Historia América*, Vol. 32, 2006, pp. 103-126.

León, Antonio García, de. La Malla Inconclusa. Veracruz y Los Circuitos Comerciales Lusitanos En La Primera Mitad Del Siglo XVI, Antonio Ibarra y Guillermina del Valle (coords.), *Redes sociales e instituciones comerciales en el imperio español, siglos XVII a XIX*. Instituto Mora/Facultad de Economía Universidad Nacional Autónoma de México. México, 2007.

Linschoten, Jan Huyghen van (Iwao, Seiichi Translation). *Tohoshokokuki*, Iwanamishoten, Tokyo, 1973.

Lasucuain García, Ana Rita Valero, de. *La Ciudad de México-Tenochtitlán Su Primera Traza 1524-1534*, México Editorial JUS, 1991.

Lorenzo Sanz, Eufemio. *Comercio de España con América en la época de Felipe II*, 2 Vols., Valladolid, Diputación Provincial, 1979.

Loureiro, Rui Manuel. *Pelos Mares da China*, CTT Correios, Lisbon, 1999.

McAlister, Lyle N. *Spain and Portugal in the New World*, University of Minnesota Press, Minneapolis,1985.

Meilink-Roelsofsz, M. A. P. *Asian Trade and European Influence in the Indonesian Achipelago Between 1500 and about 1630*, Martinus Nijhoff, The Hague, 1962.

Mikami, Tsugio. *Boueki Tojishi Kenkyu*, Vol. 3, Chuokouronbijutsushuppan,Tokyo, 1988.

Molina, Antonio M. *America en Filipinas*, Colecciones MAPFRE, Madrid, 1992.

Morga, Antonio. *Sucesos de las Islas Filipinas*, Edicion critica y comentada y studio preliminary de Francisca Perujo, Fondo de Cultura Economica, Mexico, 2007.

Mier, Lucía y Rocha, Terán, *La primera traza de la ciudad de México 1524-1535 Tomo I*, Fondo Cultura Económica, México, D. F., 2005.

Miyata, Etsuko, *El Comercio de Porcelanas Orientales en el Galeón de Manila durante los siglos XVI-XVII*,Trabajo de Investigación Tutelado, Universidad de Santiago de Compostela, 2007.

Miyata Etsuko, Spain Galicia Chiho Shutudo no Chugokutoji, *Kindaikouko*, 60, 12-16, 2008.

Miyata Etsuko, Chinese ceramics from Spain: their significance in the 16th and 17th century Atlantic coastal trade, *52 Congreso Internacional de Americanistas 2009*, Online distribution.

Miyata Etsuko, Tojiboueki ni Okeru Taiheiyou Routo-Asia to Latin America wo Musubu Touyoutoji no Nagare, *Bijutsu ni Kansuru Chosa Kenkyu Josei no Houkoku*, Kajima Bijutsuzaidan, Tokyo, 2005.

Misugi, Takatoshi. *Umi no Silkroad*, Shincho Sensho, Tokyo, 1984.

Nishida, Hiroko and Degawa Tetsurou, *Minmatsu Shinsho no Minyou*, Heibonsha, Tokyo, 1997.

Nogami, Takenori. Macao Shutsudo no Hizen Jiki, *Kindaikouko*, 50:7-11, 2005.

Nogami, Takenori, Orogo, Alfredo B., Cuevas, Nida T., Tanaka, Kazuhiko. 'Hizen Wares Excavated in Intramuros', *Kindaikouko*, 51:5-9, 2005.

Nakajima, Gakusho. Portugal jin no Nihon hatsuraikou to Higasiasia kaiikikoueki, *The Shien* 142, Kyushu University, 2005, pp. 33-72.

Nakajima, Gakusho (ed.) *Namban-Koumo-Toujin:16-17seikino Higasi Asia Kaiiki*, Shbunkaku Shuppan, Tokyo, 2013.

Ohashi, Koji. 16-17 Seiki ni okeru Nihon Shutsudo no Chugoku Tojiki ni tsuite, *Okazaki Takashi sensei Taikan Kinen Ronbunshuu: Higashi Asia no Kouko to Rekishi*, Tokyo, 1987, p. 611.

Ohashi, Koji. 'A study of the ceramic trade at the Tirtayasa site, Banten, Indonesia: The strategic point through the Ocean Silk Road, *Bulletin of the Reasearch Center for Silk Roadology*, Vol. 20, Nara, 2004.

Okamoto, Yoshitomo. *Jurokuseiki Nichiou Koutsushi no Kenkyu*, Hara Shobo, Tokyo, 1974.

Ollé, Manel. 'Interacción y Conflicto en el Parián de Manila', *Illes i Imperis:* núm.10/11, primavera, Departament d›Humanitats/Universitat Pompeu Fabra, Barcelona, 2008.

Ono, Masatoshi. 15-16 Seiki no Sometesuke Wan, Sara no Bunrui, *Trade Ceramic Studies*, No.2, Trade Ceramic Society, Tokyo, 1982, pp. 71-87.

Pierce, Donna and Otsuka, Ronald. *Asia & Spanish America Trans-Pacific Artistic & Cultural Echange,1500-1850*, Oklahoma University Press, Denver, 2009.

Pike, Ruth. *Aristocrat and Traders. Sevillan society in the sixteenth century*, Ithaca, 1972.

Pike, Ruth. *Enterprise and adventure. The Genoese in Seville and the Opening of the New World*, New York, Cornell University Press, 1966.

Pinto de Matos, Maria Antonia. *A Casa das Porcelanas*, Cerámica Chinesa da Casa-Museu Dr. Anastasio Gonçalves, Lisboa, 1996.

Pijl-Keter, P. L., van der. *The Ceramic Load of the Witte Leeuw*, Amsterdam, Rijksmuseum, 1982.

Pires, Tomé (Shigeru Ikuta Translation). *Tohoshokokuki*, Daikokaisousho, 1973.

Rees, Peter. *Transportes y comercio entre México y Veracruz, 1519-1910*, México: Sepsetentas, 1976.

Reid, Anthony. *Southeast Asia in the Age of Commerce 1450-1680, Volume Two: Expansion and Crisis*, Yale University Press, New Haven and London, 1993.

Rinaldi, Maura. *Kraak Porcelain: A Moment in the History of Trade*, London, Bamboo Publishing, 1990.

Robertson, James Alexander and Blair, Emma Helen (eds). *The Philippine Islands 1492-1898*, Vol. VII, Cleveland, 1903-1909.

Rodríguez, Manuel Xusto. *Hasta el Confín del Mundo: Diálogos entre Santiago y el Mar*, Vigo: Gráficas Varona, 2004.

Rubial García, Antonio (coordinador). *Historia de la Vida Cotidiana en México, Tomo II: La Ciudad Barroca*, El Colegio de México y Fondo de Cultura Económica, México, 2005.

Sakuma, Shigeo. *Keitokuchin Yoghyoshikenkyu*, Dai-ichi Shobo, Tokyo, 1999.

Shangraw, Clarence and Von der Porten, Edward, *The Drake and Ceremeño Expeditions, Chinese Porcelains at Drake's Bay, California, 1575 and 1597*, Santa Rosa Junior College and Drake Navigators Guild, 1981.

Shaw, Carlos Martínez (ed.), *El Pacífico español de Magallanes a Malaspina*, Madrid, Ministerio de Asuntos Exteriores y Lunwerg Editores, 1988.

Sheaf, Kolin and Kilburn Richard, *The Hatcher Porcelain Cargoes*, London, Phaidon-Christies, 1989.

Shurz, Lytle William. *The Manila Galleon*, New York, Dutton & Co., 1959.

Shurz, Lytle William. 'Acapulco and the Manila Galleon', *Southwestern Historical Quarterly*, XXII, 1918a, pp. 107-126.

Shurz, Lytle William. 'The Voyage of the Manila Galleon from Acapulco to Manila', *Hispanic American Historical Review,* II: 4 (1919), pp. 632-638.

Shurz, Lytle William. The Spanish Lake, *Hispanic American Historical Review*, 2 (1922), pp. 491-508.

Souza, George Bryan. The Survival of Empire-Portuguese Trade and Society in China and the South China Sea 1630-1754,Cambridge University Press, Cambridge,1986

Sousa, Lucío de. 'Legal and Clandestine Trade in the History of Early Macao: Captain Landeiro, the Jewish King of the Portuguese from Macao', *Kanagawa Prefectural Institute of Language and Culture Studies*, 2012, pp. 49-63.

Taín Guzmán, Miguel. *La Casa del Deán de Santiago de Compostela, la Coruña*, Diputación de la Coruña, 2004.

Takase, Koichiro. Kirishitan Kyoukai no Boueki Katsudou-Tokuni Kiito Igai no Shouhin nitsuite, *The Socio-Economic History Society* 43(1), Tokyo, 1977, pp. 54-72.

Tateishi, Hirotaka (ed.), *Daikoukai no Jidai-Spain to Shintairiku*, Doubunkan, Tokyo, 1998.

TePaske, John J. 'New World Silver, Castile, and the Far East (1590-1750)', in John Rochards (ed.), *Precious Metals in the Later Medieval and Early Modern World*, Carolina Academic Press, Durham, 1982.

Texeira, Manuel. *The Japanese in Macau*, Instituto Cultural de Macau, Macao, 1990.

Torre Revello, José. 'Merchandise brought to America by the Spaniards', *Hispanic American Historical Review* 23, 1943.

Ueda, Hideo. 16 Seiki Matsu kara 17 Seiki Zenhan ni okeru Chugokusei Sometsuke Wan, Sara no Bunrui to Hennen e no Yosatsu, *Kansai Koukogaku Kenkyuka*i, 1982, pp. 56-78.

Ueda, Makoto.*Umi to Teikoku, Chugoku no Rekishi Ming Xing Jidai*, Kodansha, Tokyo, 2005.

Valero Garcia Lascurain de, Ana Rita. *La Ciudad de México-Tenochtitlán: su primera traza (1524-1534)*, Editorial JUS Mexico, 1991.

Volker, T. *Porcelain and the Dutch East India Company*, E. J. Brill, Leiden, 1971.

Von Der Porten, Edward P. 'Drake and Cermeno in California: Sixteenth Century Chinese Ceramics', *Historical Archaeology*, 1972, pp. 1-20.

Wickberg, Edgar. *The Chinese in Philippine Life 1850-1898*, Ateneo de Manila University Press, 2000.

Yuste, Carmen. *El comercio de Nueva España con Filipinas 1590-1785*, Instituto Nacional de Antropología e Historia, México, 1984.

Yuste, Carmen. *Emporios Transpacíficos-Comerciantes Mexicanos en Manila 1710-1815*, Universidad Nacional Autónoma de México, 2007.

Whitehouse, David. Chinese Porcelain in Medieval Europe, *Bollettino d'Arte*, 1966, pp. 63-73.

Catalogues

Azul e Branco da China, Porcelana Ao Tempo dos Descobrimentos, Colecçao Amaral Cabral, Lisboa, 1997.

The Binh Thuan Shipwreck, Christie's Australia, 2004.

The Fort San Sebastian Wreck: A 16th-century Portuguese Wreck off the Island of Mozambique, Christie's, Amsterdam, 2004.

Chinese Ceramics Found in the Philippines, Hagi Uragami Museum, 2000.

El Galeón de Manila, Ministerio de Educación, Cultura y Deporte, Madrid, 2000.

Nossa Senhora dos Mártires: The Last Voyage, Verbo, Lisbon, 1998.

Rande 1702: arde o mar, Vigo: Museo do Mar de Galicia, 2002.

El San Diego; un Tesoro bajo el mar, 1995.

Saga of the San Diego, National Museum of the Philippines, 1993.

Kouro Asia e-Sakoku Zenya no Touzai Kouryu, Tabaco to Shio no Hakubutsukan, Tokyo, 1998.

Talaveras de Puebla Ceramica colonial mexicana Siglos XVII a XXI, Museu de Ceramica de Barcelona, 2007.

Appendix 1
AGN Contratación 1795-1802

Ships and cargos	Despatching port/Year	Contents of boxes	Sender	Resident of	Reciever	Resident of
San Bartolomé	Veracruz 1591					
silver						
dye material						
reales						
hide						
1 box of balsam						
1 box		2 dozens of Chinese ceramics, plates, bowls				
		10 pieces of Chinese ceramics				
		root of Michoacan (dye material)			Juan Lorenzo	Sevilla
silverwares						
1 box of ceramics			Graviel de Balmaceda	Mexico	Lucas Valorado	Sevilla
La Concepcion	Veracruz					
silver						
reales						
gold						
hide						
cochineal						
indigo						
Santa Susana	Veracruz					
gold						
silver						
reales						
Nuestra Señora de la Concepción	Veracruz					
gold						
silver						
reales						

APPENDIX 1 - AGN CONTRATACIÓN 1795-1802

musk	product				
	1 box	1 bed with decorated with taffeta from China			
		2 others			
		3 pairs of curtains for church altars			
		1 small canopy			
		2 clothes of Our Majesty embroidered with silk			
		4 pillow and 4 pincushions embroidered with silk			
		3 scarfs for covering caliz embroidered with silk			
		1 bag embroidered for corporals			
		6 pieces of white envelopes			
		12 canudos of silk, 6 white and 6 brown			
		1 dutch (textile) of silk			
		3 pieces of white stumpesos			
		3 pieces of painted stumpesos			
		1 pieces of taffeta embroidered with blue thread			
		5 ropes of silk from Japan			
		a small amount of laces, small amount of twisted colored silk	Simón Rodríguez	Mexico	Rui Fernandez	Sevilla
	1 box	2 series (of books?) from China gold decorated			

Portuguese Intervention in the Manila Galleon Trade

Container	Item	Owner		Recipient	
	100 fans decorated with gold in 2 caetas				
	3 more (fans) decorated with gold				
	3 palo de Aquila				
1 box	some plates from China				
	2 clothes decorated with gold				
	hide				
	11 colored cloth				
	12 colored shelves	Francisco Conde	Mexico	Francisco de Torres	Sevilla
2 boxes	silk (10 arrobas and 14 pounds)				
4 boxes	box with key decorated with gold from China				
	shawl of feather				
	4 pieces of damask				
	1 piece of white satin				
	a small amount of thread				
	other meticulousness of feather				
	3 bottles of oil				
	figure of Christ	Antonio Maldonado			
	a painting of San José of feather				
1 suitcase with things from China	1 painting of N.Señora de Entales				
	things with feather				
	12 spoons of pearl from China				
	1 cushion of white feather				
	2 protections of velvet				
	6 pieces of golden cloth				

APPENDIX 1 - AGN CONTRATACIÓN 1795-1802

	1 piece of yellow, colored and black damask from China			
	1 piece of green damask from China			
	1 piece of white damask from China			
	37 pieces of pajaras (unidentified product) from China			
	10 pieces of damask and tafettas y white satins of different colors			
	7 almoyzales (unidentified product) from China			
	1 canopy from Michoacan			
	1 cushion of feather	Gerónimo Perez Aparicio (Dr. Pº Farfán del consejo de SM: passenger)	Veracruz	Hernando Carmona
Nuestra Señora de la Concepción				
silver			Veracruz	Domingo de Corcuera
reales				
530 pesos of tipus (unidentified product)				
San Juan Bautista			Veracruz	
cochineal				
silk	120 pounds of raw silk	Gerónimo Perez Aparicio		
hide				
indigo				
5 boxes of things from China		Gerónimo Perez Aparicio		

Buen Jesús surto	Veracruz					
reales						
gold						
silver						
hide						
indigo						
2 boxes of Michoacan cochineal						
2 boxes of Chinese ceramics			Pedro de Yrala (Pedro de Carmona)	Veracruz	Ines de Santa Ana (mother of Pedro Carmona)	Sevilla
1 box of raw silk		116 pounds	Joan de Villa (Francisco de Paz)		Andres Franco	Sevilla
1 box		212 pieces of large and small Chinese ceramics	licensed for Antonio de Paz		Pedro Moya de Contreras	Mexico
Santa María de Begonia	Veracruz 1593-1595					
sugar						
cochineal						
hide						
indigo						
1 box of raw silk		230 pounds				
things of feather			Baltasar de Valdes	Jerez		
gold						
reales						
silver						
silverwares						
San Juan de la Magdalena	Veracruz					
hide						
palo (plant for dye material)						

APPENDIX 1 - AGN CONTRATACIÓN 1795-1802

cochineal					
Nuestra Señora de Ayuda	Veracruz				
hide					
cochineal					
palo					
Nuestra Señora de Encarnación	San Cristobal de Havana				
hide					
canafiola? (unidentified product)		5 quintales (1quintal=46kilos)			
anime (unidentified product)					
San Palero	Veracruz (surto)				
silver					
gold					
reales					
indigo					
hide					
cochineal					
San Juan de Bautista	Veracruz				
palo					
cochineal					
San Andrés	Mérida, Yucatán				
reales		365pesos			
wine		16 pipas			
breaqueme (unidentified product)		5 quintales	Pedro de Tapía	Sevilla	

Portuguese Intervention in the Manila Galleon Trade

Ship without a name	1595				
silver gold					
2 boxes of silk					
5 boxes of silk		766 pounds		Rodrigo de León	Sevilla
hide					
indigo					
1 boxes		98 large and small pieces of Chinese ceramics	Alonso de Velorado	Mexico	Isabel Hurtado
		blanket of white satin from China			
		cup with medium saucer decorated with gold			Padre Maestro Fray Joan Ramírez
chocolate					
1 box		some portraits			
		1 pound of thread			
		4 pieces of Chinese game			
		2 pairs of colored rriendas			
1 box		1 sack with 3 colored suitcase			
		208 catties	Baltasar de Alca(Joán de León Castilla)	Veracruz	Fernando de Medina Campo
		1 pan			
		1 jar			
		cup with gold decoration on the foot			
		4 mediam plates			
		1 large and small spoon			
		1 salt shaker			
		3 dozens of Chinese plates and bowls	Alonso de Arroyo (pasajero)		
		7 dozens of damasks from China			

Appendix 1 - AGN Contratación 1795-1802

Item	Ship/Date	Description	Sender	Origin	Recipient	Destination
		some tablecloths				
		6 rods of blue damasks from China				
		1 piece of white satin from China				
		2 pounds of pita (unidentified material)				
		1 table of Chinese chess				
		1 portrait of San Francisco				
		1 figure of small Christ	Alvaro de Baena (Gerónimo Bonifas)	Veracruz	Gaspar Espinas Bonifaz	
algalia						
1 box		oil				
silk		90 pounds of raw silk from China				
3 boxes of silk		16 arrobas (1 arroba = 30 pounds)				
gold						
Nuestra Señora de Rosario	San Cristobal de Habana 1594					
silver						
raw silk			Alonso Martín Merader	Mexico	Andrés Franco	Sevilla
Nuestra Señora de la Candelería	1595					
hide						
carca						
1 box		golden crown from China	Manila Joan Sarmento	Sevilla		
indigo		3 ornaments of silk				
cochineal						
chocolate						

			Pedro Joseph Carboneli		
1 box of silk					
Nuestra Señora de Rosario	San Juan del Puerto de Cavallo (Honduras)				
palo of Brazil (dye material)					
wax					
indigo					
ink of indigo					
Xarca (unidentified material)					
balsam					
hide					
carca (unidentified item)					
10 bottles of Peruleras (unidentified material)					
ginger					
Espiritu Santo					
hide					
cochineal					
1 box		97 catties of silk		Sebastian de Barrada	Cartagena	Joan de Chavez
4 boxes of medicine of Michoacan					
Santa Ana	San Juan del Puerto de Cavallo (Honduras)				
indigo					
bottle of balsam					
hide					
xarca					

Nuestra Señora de Esperanza		
gold		2000 pesos of gold común equivalent to 8 reales
reales		300 pesos de a 8 reales cada peso
reales		227 pesos de a 8 reales cada peso
reales		400 pesos de a 8 reales cada peso
silver		1250 pesos de oro común
		143 marcas and 2 ounce
		130 marks
		124 marks
		149 marks
		124 marks and 2 ounce
		144 marks and 2 ounce
		114 marks
		107 marks and 3 ounce
		1073 pesos and 1 tomin in silver
		920 pesos and 5 tomines
		888 pesos and 2 tomines
		838 pesos and 4 tomines
		89 marks and 2 ounce
		1056 pesos 7 tomines
		137 marks and 3 tomines
silver from Peru		2 barras (142 marks y 6 ounce)
silver		265 marks

Portuguese Intervention in the Manila Galleon Trade

	1000 pesos of common gold	250 pesos in reals	1007 pesos and 7 tomines (=reales)																								
reales	reales	silver	cochineal	reales	reales	silver	reales	silver	reales	reales	silver	reales	silver	silver	silver	reales	silver	silver	silver	reales	reales	reales	reales	reales	silver	reales	silver	reales	reales

Appendix 1 - AGN Contratación 1795-1802

reales	reales	reales	reales	reales	silver	reales	reales	hide	silver	silver	silver	cochineal	reales	silver	silver	reales	reales	reales	cochineal	reales	reales	cochineal	silver	silver	silver	silver	silver	silver	silver	reales

silver	proceso	silver	reales	reales	silver	silver wares	reales	reales	silver	reales	reales	reales	reales	reales	reales	reales	reales	2 boxes with presents	silver	silver	silver	silver	silver	silver	reales	reales	reales	reales	reales	reales

reales	silver	reales	silver	reales	silver	silver	silver	reales	silver	reales	reales	reales	silver	reales	silver	reales	silver	reales	cochineal		hide	reales	cochineal	reales	silver	silver	reales	2 boxes of silk	127 pounds

127 pounds	cochineal			
20 arrobas	cochineal			
5 arroba	cochineal			
10 arroba	cochineal			
20 arrobas	cochineal			
10 arrobas	cochineal			
20 arrobas	cochineal			
5 arroba	cochineal			
20 arrobas	cochineal			
370 pounds in total	3 boxes of silk			
30 arrobas	cochineal			
8 arrobas	indigo			
14 bowls with gold decoration	3 boxes			
12 saltshaker				
10 plates				
4 bowls				
1 large round porcelan bowl				
12 plates				
9 porcelains each with handles				
1 bottle of oil, small box		Juan y Lucas Perez de Ribera	Mexico	
4 barrels of balsam				
4 barriles of oil				
4 boxes of tecomachan (unidentified product)				
2 arrobas of incense				

Appendix 1 - AGN Contratación 1795-1802

hide	cochineal	silk	indigo	cochineal	silk	silk	gold	hide	hide	hide	hide	perlas	reales	reales	reales	reales	reales	reales	reales	reales	reales	pearls	Santiago el Mayor	cochineal	indigo	cochineal
					Pedro Deyra (in the name of Pº Lopez de Tapia)	Pedro Deyra(in the name of Alonso Rassa)																	Havana			
					Veracruz																		1598			

Portuguese Intervention in the Manila Galleon Trade

wheat				
silver				
hide				
3 boxes	incense			
	root of Michoacan			
San Augustín				
cochineal				
indigo				
silver				
magdalen				
chickpea				
cochineal				
San Buenaventura	puerto de Santa Povadela			
silk	460 pounds			
hide				
cochineal				
indigo				
palo				
1 box of Chinese ceramics		Pedro de Lopez de Tapia	Veracruz	
San Francisco				
hide				
cochineal				
moranos of silk				
palo				
bull				
1 box	4 picos of damasks from China			

Appendix 1 - AGN Contratación 1795-1802

	root of Michoacan			
Gold				
2 boxes	bottle of oil			
	incense			
box of various things	1 pico of blue satin from China			
	3 picos of damasks from China			
	6 plates of ebony from China			
	2 bowls of clay from China			
	other porcelain with colors			
	Porcelains			
	2 small colored plates from China			
	6 small bowls decorated with gold and colors			
	10 large bowls decorated with gold and other colors			
	a dozen and a half of fans	Veracruz	Diego de Herrera	Sevilla
1 sewn box	9 dozens of Chinese ceramics			
	cooking pan			
	some rosaries			
	strings			
	172 pesos of gold		Francisco de Salcedo	Sevilla
1 box	things from China	Mexico	Domingo de Corcuera	Sevilla
3 boxes	things from China		Francico de Tello	Sevilla
cacao				
1 box	things from China		Antonio Olivera	Sevilla

Nuestra Señora del Rosario	1601							
hide								
silver								
cochineal								
indigo								
indigo from Guatemala								
cochineal								
palo								
reales								
3 boxes of silk								
Espiritu Santo								
oro comun								
silver								
palo								
hide								
3 boxes of silk	11 arrobas of raw silk				Juan Faraz			Sevilla
	4 arrobas of floral silk							
1 box of raw silk	150 pounds of raw silk			Alonso Díaz de la Varela		México	Juan Rodríguez	Sevilla
2 boxes of raw silk	300 pounds raw silk from China						Juan Rodríguez	Sevilla
1 box of raw silk	100 pounds of raw silk from China						Bernardo Alfonso	Sevilla
silk	4 arrobas and 19 pounds			Diego Lopez de Montalvan		México	Gerónimo Lopez de Cabrera	Sevilla
gold jewelry								
1 box of raw silk							Juan Bautista	
1 box of raw silk								
2 boxes of raw silk				Francisco de Nobili?, Miguel Gerónimo		México	Abaetazar Espino	